Source Mechanism and Seismotectonics

Edited by
Agustín Udías
Elisa Buforn

Springer Basel AG

Reprint from Pure and Applied Geophysics
(PAGEOPH), Volume 136 (1991), No. 4

Editors' addresses:

Agustín Udías
Elisa Buforn
Universidad Complutense
Facultad de Ciencias Físicas
Departamento de Geofísica
Ciudad Universitaria
E–28040 Madrid
Spain

Deutsche Bibliothek Cataloging-in-Publication Data

Source Mechanics and seismotectonics / ed. by Augustín Udías;
Elisa Buforn. – Basel ; Boston ; Berlin : Birkhäuser, 1991
   ISBN 978-3-0348-9718-1   ISBN 978-3-0348-8654-3 (eBook)
   DOI 10.1007/978-3-0348-8654-3
NE: Udías, Augustín [Hrsg.]

© 1991 Springer Basel AG
Originally published by Birkhäuser Verlag Basel in 1991
Softcover reprint of the hardcover 1st edition 1991

ISBN 978-3-0348-9718-1

Contents

369 Preface

371 Introduction, *R. Madariaga and A. Udías*

375 Intermediate and deep earthquakes in Spain, *E. Buforn, A. Udías and R. Madariaga*

395 Spanish national strong motion network. Recording of the Huelva earthquake of 20 December, 1989, *E. Carreño, J. Rueda, C. López Casado, J. Galán and J. A. Peláez*

405 Regional focal mechanisms for earthquakes in the Aegean area, *B. Papazachos, A. Kiratzi and E. Papadimitriou*

421 Rates of crustal deformation in the North Aegean trough-North Anatolian fault deduced from seismicity, *A. A. Kiratzi*

433 Regional stresses along the Eurasia-Africa plate boundary derived from focal mechanisms of large earthquakes, *A. Udías and E. Buforn*

449 Focal mechanisms of intraplate earthquakes in Bolivia, South America, *A. Vega and E. Buforn*

459 Partial breaking of a mature seismic gap: The 1987 earthquakes in New Britain, *R. Dmowska, L. C. Lovison-Golob and J. J. Durek*

479 Size of earthquakes in Southern Mexico from indirect methods, *J. A. Canas*

499 Numerical simulation of the earthquake generation process, *M. Radulian, C.-I. Trifu and F. O. Cărbunar*

515 Intermagnitude relationships and asperity statistics, *A. A. Gusev*

529 Complete synthetic seismograms for high-frequency multimode *SH*-waves, *N. Florsch, D. Fäh, P. Suhadolc and G. F. Panza*

561 Body-wave dispersion: Measurement and interpretation, *A. M. Correig*

PAGEOPH, Vol. 136, No. 4 (1991)

0033–4553/91/040369–2$1.50 + 0.20/0
© 1991 Birkhäuser Verlag, Basel

Preface

This issue of Pure and Applied Geophysics contains papers presented at the Symposium on Earthquake Source Mechanism and Seismotectonics that was convened by the Subcommission on Physics of Earthquakes Sources of the European Seismological Commission and sponsored by the Universidad Complutense de Madrid, Instituto Geográfico Nacional, Madrid and the Institut de Physique du Globe, Paris and held from 28 to 30 May 1990 in El Escorial, Madrid, Spain. This meeting is a continuation of that organized by the same Subcommission in 1979 in Mogilany, Poland. Some papers were presented at the scientific session of the same Subcommission during the XXII General Assembly of the European Seismological Commission, 17 to 22 September 1990, Barcelona.

This issue contains 12 papers on topics broadly related with the problem of the source mechanism of earthquakes and its applications to the seismotectonics of seismic active regions. Most of the papers belong to the applied aspects of the problem rather than to purely theoretical questions. Some of them study the source mechanism of earthquakes in some selected areas, such as the Mediterranean region, South Spain, Aegean area, central Bolivia and New Britain, and relate them to the tectonic processes active in each region. Theoretical and applied aspects are treated on the generation of synthetic seismograms, intermagnitude relationships, numerical simulation of earthquake processes and determination of earthquake size from indirect measurements. Observational and instrumental aspects related to the subject are presented on the Spanish strong motion network and the Mexican seismic warning system.

Unfortunately, not all papers presented in the symposium are published in their entirety in this issue. The symposium included interesting discussions on theoretical and applied aspects of the problem of earthquake source mechanism. Tectonic applications of the results from earthquake mechanisms to different areas, from the viewpoint of geology and seismology, generated interesting discussions.

The editors are grateful to the institutions which sponsored this symposium. Appreciation is expressed to the editorial board of Pure and Applied Geophysics for providing this opportunity to edit and publish the papers of the symposium. Specially, sincere thanks are extended to Dr. R. Dmowska, and Dr. E. Okal for their valuable assistance during the entire process of preparing this publication. Financial help in the organization of the symposium from the Universidad

Complutense and the Dirección General de Investigación Científica y Técnica of the Ministerio de Educación y Ciencia of Spain is acknowledged.

A. Udías
E. Buforn
Guest Editors

PAGEOPH, Vol. 136, No. 4 (1991)

0033–4553/91/040371–4$1.50 + 0.20/0

Introduction

RAÚL MADARIAGA[1] and AGUSTÍN UDÍAS[2]

The relationship between earthquake source mechanism and plate tectonics was demonstrated almost 20 years ago by ISACKS, OLIVER and SYKES (1968) in their classical work on the relationship between seismic activity and plate tectonics. Since then, earthquakes have been extensively used as tracers of seismic activity, and source mechanisms have served to determine the geometry and kinematics of plate motion. The main features of the seismicity of classical plate boundaries, subduction zones, mid-ocean ridges and major transform zones are now relatively well-understood. Progress in understanding source mechanisms in more complex tectonic regions, particularly in diffuse boundaries, has been much slower because of their generally weaker seismicity. The use of worldwide networks for the determination of source plane solutions in these regions is limited. This condition prevailed until late 70's in the Mediterranean region and its surrounding areas. The seismicity of this region is directly related to the relative motion between the African and the Eurasian plates, but instead of being distributed along a well-delimited narrow band, the activity of this boundary is diffuse and spread over a large area several hundred km wide. Source mechanisms in the Mediterranean were determined very early by RITSEMA and colleagues employing the data that was available before the installation of the WWSSN network. Many of those mechanisms were not very well constrained. The relationship between seismicity and plate tectonics in the area was first discussed by MACKENZIE (1972), who redetermined most of the source mechanisms that were well constrained from standard long-period far-field data. These data, as well as those obtained by many other researchers, established several important characteristics of the mechanism and distribution of the larger earthquakes of the region and their relationship to its main geological features: active subduction along the Aegean trench, end of subduction in the Calabrian arc, extension inside the Aegean plate and along the Apennines, complex seismicity

[1] Laboratoire de Sismologie, Institut de Physique du Globe and Université Paris 7, 4 Place Jussieu, Tour 14, 75252 Paris Cedex 05, France.
[2] Departamento de Geofísica, Universidad Complutense de Madrid, Ciudad Universitaria, 24040 Madrid, Spain.

along the eastern Adriatic Sea, diffuse compression in northern Algeria and a complex transform boundary along the Azores Gibraltar area.

Similar to other intraplate regions, a more detailed understanding of the seismotectonics of the Mediterranean awaited the development of regional seismic networks equipped with homogeneous instruments. Only those kinds of instruments can produce reliable data for the determination of the source mechanism of smaller but important regional shocks. Telemetered networks have been installed or are in the process of installation in many Mediterranean countries. The data that they provide are changing the general perception of the seismicity of the region and will make possible increasingly detailed studies of seismotectonics. Several communications in this symposium report on the most recent results coming from the data obtained from these improved seismic networks. Illustrative of this is the paper presented by BUFORN et al. who report on the mechanism of several intermediate and deep events that occurred below southern Spain. These earthquakes could be studied thanks to the telemetered digital network installed in Spain by I.G.N. CARREÑO et al. present the main characteristics of this network and present the data obtained after a medium-sized event that occurred in 1989 near Huelva in southern Spain. Greece is by far the most active region in the Mediterranean, several shocks of magnitude 5 or greater occur there every year. PAPAZACHOS et al. and KIRATZKI present some recent data on seismicity and source mechanism from the Aegean and the Northern Anatolian fault. A very active region in Eastern Europe is the Carpathian Arc where large intermediate depth earthquakes of magnitudes reaching 7 have produced great damage in Romania. RADULIAN et al. review some of the work they have done on the seismicity of the region and propose a model for the time and space evolution of this seismicity, based on recent work on the percolation of seismicity clusters. A synthesis of currently available data on source mechanisms and the stress field along the Africa-Eurasia diffuse plate boundary is provided by the paper of UDIAS and BUFORN.

Recent studies of seismicity and focal mechanism in the Mediterranean have also clearly put the finger on the nearly obvious problem that it is impossible to understand the origin of seismic activity without a multidisciplinary approach involving earth scientists of different origins. The strong interaction between seismologists, tectonicists, geodesists and other geophysicists who studied several important events that occurred in the early 1980s, led to a major revision of the approach to study large Mediterranean earthquakes in the field. The main outcome of these studies, an integrated vision of the complete seismic process, has opened the way to new approaches towards the understanding of seismotectonics. Some new paleoseismological data, recently obtained from the El Asnam region in Algeria, and from the Irpinia area in Central Italy, poses a number of problems concerning the return period of large earthquakes in these sites. Future work to be carried out in Greece should provide new insight on the seismic cycle of the most active faults in the region. Recent studies of destructive earthquakes in Peloponnesus, Armenia, and Georgia have largely profited from the experience acquired in the study of the large events of the early 1980s. The Mediterranean region is not only the site of shallow dispersed seismic activity, it is also the site of localized intermediate depth seismic activity, and of some rare deeper events. Active subduction is clearly at the origin of Greek intermediate depth seismicity, while the deep activity under Calabria and the Vrancea region in Romania is clearly related to downgoing slabs that have long remained active after the arrest of subduction. The origin of the intermediate and very deep seismicity below Spain is a considerably more complex problem. Several possible explanations of the intermediate depth activity to about 110 km have been proposed, but the origin of the deep (650 km) events studied by BURFORN *et al.* remains a mystery since this is the only source of deep activity that is not related in an obvious way to a subduction zone.

The problem of seismic source mechanism and seismotectonics cannot be completely separated from that of the mechanical origin of earthquakes. Recent work on this subject has revealed the intimate relation between fault geometry, as observed by field geologists, and the results of inversion of source process from seismic data in the far-field or in the near-field. In our opinion, the study of seismic mechanism is approaching a crucial point when broad-band instruments sitting next to faults will be available to complement the classical far-field long and intermediate period seismograms. In fact, most of what we have learned about source mechanics has come from the painstaking modelling of large seismic events under Island Arcs. These events are unfortunately inaccessible to direct observation so that source models cannot be refined beyond a certain point. In particular, it is impossible to obtain maps of the faults that have been activated during any of these subduction zone events. The study of smaller, well-instrumented fault zones in the Mediterranean or other seismically active regions on continents should lead to a significant improvement in the knowledge of the details of the control of rupture propagation by geometrical and stress heterogeneities (barriers and asperities). Barriers, for instance, have been clearly identified along the faults activated during the Irpinia and El Asnam earthquakes of 1980. Broad-band studies of medium-sized events should provide unique data on the nature of the intermediate frequency band, 1–8 s where data is presently most lacking. It will then be possible to reconstruct a single picture of fault processes from the usual long-period far-field seismic band to the considerably more detailed accelerometric data coming from the near field. Several papers addressing these problems were presented in the conference but have been published elsewhere.

In recent times, a subject of great interest to physicists and geophysicists has been the nature of seismic activity on mature faults: faults are complex, but is their geometry fractal or does it have some characteristic lengths that we have failed to identify? Is the stress field being constantly driven to a critical point where earthquakes of any size can occur at any time as proposed by tenants of the model of self-organized criticality? Several of the problems posed by complexity were discussed during the conference. A multidisciplinary approach, incorporating methods developed by physicists to study percolation and complex systems, may prove to be very fruitful in the future. The papers by GUSEV and RADULIAN *et al.* discuss some of these points.

Finally, earthquake studies from seismic data are completely dependent on our ability to stimulate seismic wave propagation in the earth. Without information about propagation, source processes cannot be retrieved reliably from seismic data. Several papers studying the modelling and correction of seismic data were presented at the conference and are represented in this issue by the contributions of PANZA *et al.*, CANAS and CORREIG who studied the simulation and use of seismic surface wave propagation in the earth.

REFERENCES

ISACKS, B., OLIVER, J., and SYKES, L. (1968), *Seismicity and the New Global Tectonics*, J. Geophys. Res. *73*, 5855–5899.

MCKENZIE, D. (1972), *Active Tectonics of the Mediterranean Region*, Geophys. J. R. Astr. Soc. *30*, 109–185.

PAGEOPH, Vol. 136, No. 4 (1991)

0033-4553/91/040375-19$1.50 + 0.20/0

Intermediate and Deep Earthquakes in Spain

E. Buforn[1], A. Udías[1] and R. Madariaga[2]

Abstract — Recent improvements in the seismological networks on the Ibero-Maghrebian region have permitted estimation of hypocentral location and focal mechanisms for earthquakes which occurred at South Spain, Alboran Sea and northern Morocco of deep and intermediate depth, with magnitudes between 3.5 and 4.5. Intermediate depth shocks, range from 60 to 100 km, with greater concentration located between Granada and Málaga. Fault-plane solutions of 5 intermediate shocks have been determined; they present a vertical plane in NE-SW or E-W direction. Seismic moments of about 10^{15} Nm and dimensions of about 1 km have been determined from digital records of Spanish stations. *P*-wave forms are complex. This may be explained by the crustal structure near the station, discontinuities in the upper mantle and inhomogeneities near the source. Deep activity at about 650 km has only 3 shocks since 1954 (1954, 1973, 1990). Shocks are located at a very small region. Fault-plane solutions show a consistent direction of the pressure axis dipping 45° in E direction. For the 1990 shock seismic moment is 10^{16} Nm and dimensions 2.6 km. The *P*-waves are of simpler form with a single pulse. The intermediate and deep activities are not connected and no activity has been detected between 100 and 650 km. The intermediate shocks may be explained in terms of a recent subduction from Africa under Iberia in SE direction. The very deep activity must be related to a sunk detached block of lithospheric material still sufficiently cold and rigid to generate earthquakes.

Key words: Focal mechanism, intermediate earthquakes, deep earthquakes, Spain, seismicity, seismotectonics, subduction.

Introduction

The occurrence of intermediate depth and deep earthquakes ($30 < h < 300$ km) is well known in regions of Europe such as the Calabrian, Hellenic and Carpathians arcs (RITSEMA, 1969; MCKENZIE, 1972; RITSEMA, 1972; PAPAZACHOS, 1973; PAPAZACHOS and COMMINAKIS, 1977; MCKENZIE, 1978; UDÍAS, 1982; UDÍAS *et al.*, 1989), but for Spain the only information in the past was the occurrence in 1954 ($M = 7$) of a very deep shock (650 km), near Granada. There were also some indications of the presence of intermediate depth earthquakes in southern Spain, the Alboran Sea and northern Morocco (MUNUERA, 1963). In the last ten years, several authors (HATZFELD, 1978; GRIMISON and CHENG, 1986; BUFORN *et al.*, 1988) confirmed the occurrence of these intermediate shocks. The new digital

[1] Dept. de Geofísica, Universidad Complutense, Madrid, Spain.
[2] Institut de Physique du Globe, Université de Paris VII, Paris, France.

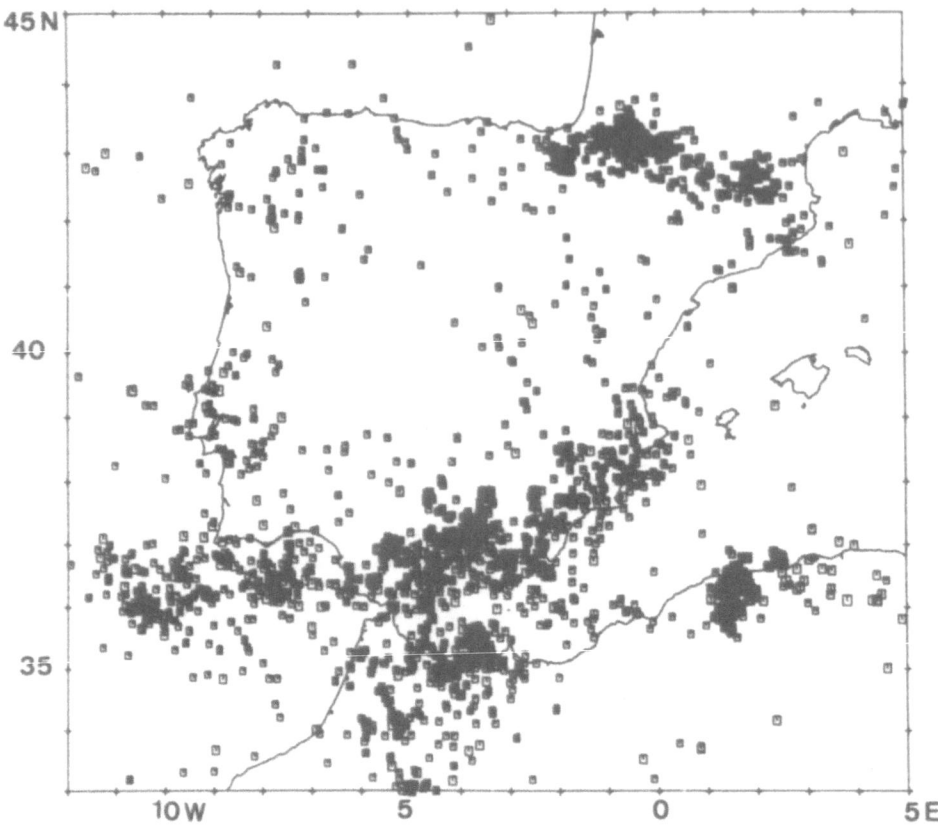

Figure 1
Epicentral distribution of earthquakes for the period 1970–1990 with magnitude equal to or greater than
3. (Seismicity Data File, Instituto Geográfico Nacional, Madrid).

seismic network installed in Spain by the Instituto Geográfico Nacional (I.G.N.) in
1979, has permitted detection with sufficient accuracy of foci with intermediate
depth between 30 and 150 km. This digital seismic network is composed of 25
vertical short period stations (MEZCUA and UDIAS, 1991). Other digital seismic
networks are installed in Morocco, Algeria and Tunisia.

In Figure 1 the seismicity for Spain is shown (Seismological Data File, I.G.N.,
Madrid) for the period 1970–1990 with magnitudes greater than or equal to 3.
Most epicenters are concentrated south of the Iberian Peninsula in a region of
deformation that marks the position of the plate boundary between Eurasia and
Africa. This contact extends from the Azores islands to the Strait of Gibraltar and
continues across southern Spain and northern Africa through northern Morocco,
Algeria and Tunisia. The two main characteristics of the seismicity in southern
Spain are the occurrence of large earthquakes ($M > 6$) separated by long-time

intervals and the continuous low magnitude activity ($M \leq 4.5$), both with shallow depths.

Intermediate Earthquakes

Intermediate depth earthquakes in this region are not frequent. Figure 2 shows the epicenters of intermediate foci ($30 < h < 150$ km) for the period 1970–1990, with depth error less than 15 km. The most important concentration is located in an arcuate zone of approximately NS direction in Granada-Málaga, west of the Alboran Sea and northern Morocco. Figure 3 shows a vertical NW-SE cross section corresponding to the profile AA′ in Figure 2 of intermediate earthquakes. Most shocks are located at a distance between 100 and 200 km from A, in the region of Granada-Málaga, and the maximum depth of the foci is 130 km. No shocks have been detected between 150 km and 650 km. At this depth, the deepest in Europe, three very deep shocks occurred, below Granada, on 29 March 1954 ($M = 7$), 30 January 1973 ($M = 4.0$) and a recent one on 8 March 1990 ($M = 4.8$).

Fault-plane Solutions

For this study, we have selected 5 earthquakes which occurred during the period 1979–1990, with intermediate depth ($60 \leq h \leq 150$ km) and magnitudes between 3.5 and 4.5. In Table 1, the hypocentral coordinates for these events are given and the locations are shown in Figure 11. In Figure 4 and Table 2 the fault-plane solutions determined from the first motion of P waves are shown. The observations correspond to short-period stations at regional distances and with most of the rays rising from the source. Data have been read by the authors from the original records. The take-off angles have been estimated from a model of the crust with a linear velocity gradient ($v_c = 5.0 + 0.0833z$) and the upper mantle with a different gradient ($v_m = 8.0 + 0.0018(z - 30)$).

Table 1

Hypocentral coordinates of earthquakes studied in this paper

No.	Date	Time	Lat.	Long.	Depth (km)	Magn.
1	29–03–1954	06–16–05.0	37.0N	3.6W	640	7.0
2	30–01–1973	02–35–59.8	36.9N	3.7W	660	4.0
3	20–06–1979	00–09–06.4	37.2N	3.5W	60	4.5
4	13–05–1986	00–19–45.6	37.6N	4.5W	90	4.5
5	27–03–1987	09–38–25.3	36.8N	4.1W	69	3.5
6	23–07–1987	11–57–27.8	35.3N	5.9W	100	3.9
7	12–12–1988	06–40–43.4	36.3N	4.6W	70	4.3
8	08–03–1990	01–37–12.3	37.0N	3.6W	637	4.8

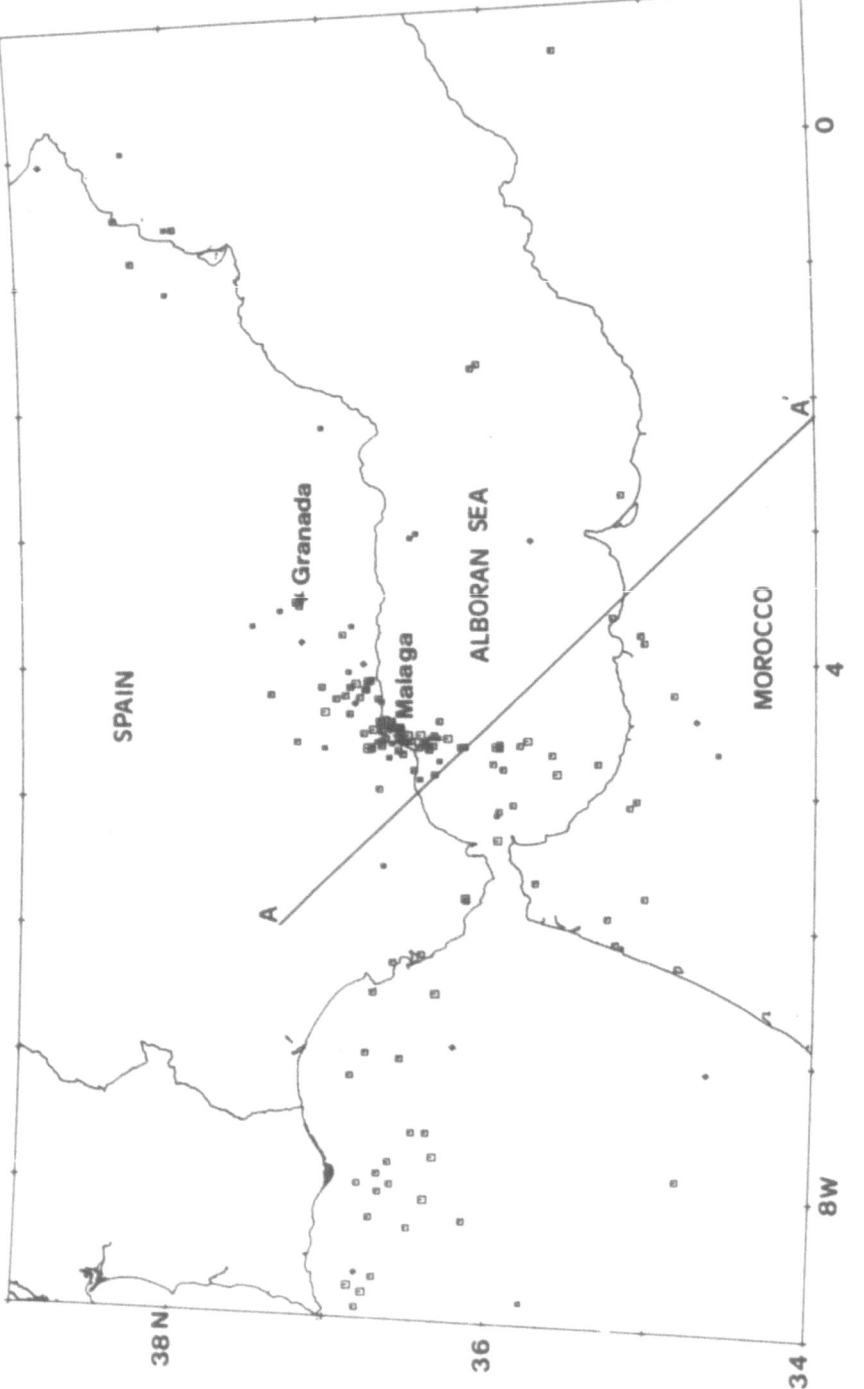

Figure 2

Distribution of foci for the period 1970–1990 with depth between 30 and 100 km (Seismicity Data File, Instituto Geográfico Nacional, Madrid).

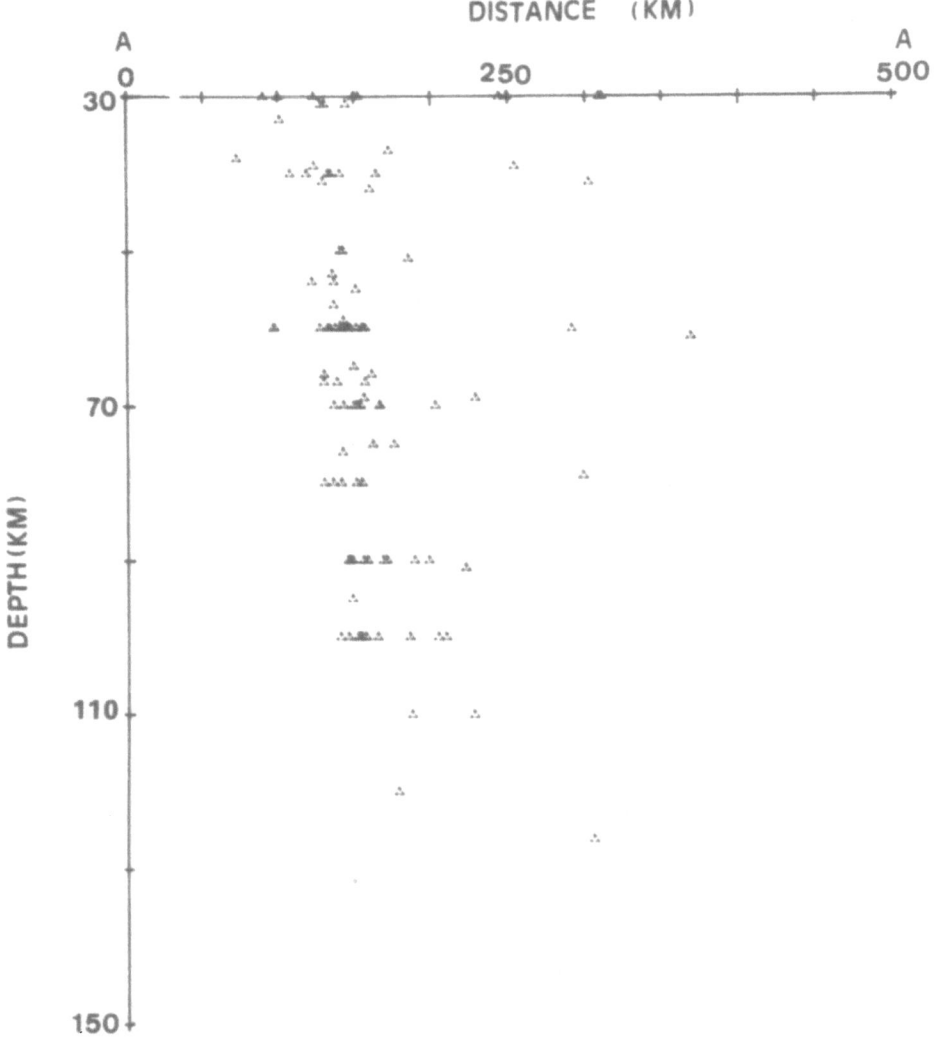

Figure 3
Vertical cross section for the foci in the period 1970–1990. The line AA′ corresponds to the one in
Figure 2.

The solutions obtained for shocks 3, 4 and 7 are similar, with a nearly vertical
plane in NE-SW direction for shocks 3 and 7 and E-W for shock 4, and the other
plane nearly horizontal except for shock 7. In all cases the pressure axis dips 45° in
NW direction. For event 5, the focal mechanism obtained has a vertical plane, in
NW-SE direction and a horizontal plane, oriented in NE-SW direction; the pressure
axis dips 45° in SE direction. The solution obtained for event 6 shows a vertical
plane oriented in N-S direction and a horizontal plane in NE-SW direction. The

Table 2

Fault-plane solutions of Spanish intermediate and deep earthquakes

No.	Axes T, P			Fault Plane			N	P
	Φ	θ	ϕ	δ	λ			
1	T:297 ± 13	55 ± 11	A: 8 ± 28	32 ± 12	−3 ± 34		89	0.9
	P: 61 ± 9	52 ± 8	B:179 ± 7	88 ± 18	−122 ± 11			
2	T:256 ± 23	68 ± 15	A:303 ± 30	37 ± 48	−153 ± 15		15	0.80
	P:139 ± 22	41 ± 16	B:191 ± 3	74 ± 12	−56 ± 49			
3	T:153 ± 10	49 ± 3	A:309 ± 10	12 ± 1	−166 ± 4		29	0.86
	P:310 ± 13	44 ± 3	B: 52 ± 24	87 ± 1	−102 ± 1			
4	T:202 ± 14	68 ± 15	A:335 ± 3	36 ± 16	−152 ± 10		27	0.89
	P:321 ± 18	41 ± 16	B: 87 ± 69	74 ± 9	−123 ± 14			
5	T:349 ± 1	31 ± 1	A:231 ± 3	16 ± 1	110 ± 3		17	0.94
	P:157 ± 1	60 ± 1	B: 72 ± 1	75 ± 1	84 ± 1			
6	T: 92 ± 6	58 ± 6	A:227 ± 13	19 ± 7	−33 ± 30		19	0.89
	P:239 ± 8	37 ± 5	B:349 ± 3	79 ± 5	−106 ± 11			
7	T:171 ± 1	60 ± 3	A:316 ± 2	50 ± 3	175 ± 4		30	0.97
	P:276 ± 2	66 ± 2	B:222 ± 1	86 ± 3	40 ± 3			
8	T:268 ± 15	73 ± 7	A: 0 ± 41	28 ± 7	−88 ± 40		40	0.95
	P: 84 ± 58	17 ± 6	B:177 ± 12	62 ± 7	−91 ± 120			

N = Number of observations P = Score

pressure axis dips 45° in NE direction. All solutions are well constrained, with most standard errors for planes and axes less than 15° (Table 2) and scores greater than 0.85.

In Figure 5, the fault plane solution for event 7 and the P-wave forms from the Spanish Digital Stations are shown. The time scale is the same for all the records but the amplitudes have been multiplied by a different factor. Stations ENIJ and EMEL; EJIF and EVAL have similar waveforms. In general the P-waves have complex forms with several arrivals which may be explained due to the small magnitudes by the crustal structure near the station, discontinuities in the upper mantle and inhomogeneities near the source.

For this region, only three events before 1976, with intermediate depth, have fault-plane solutions (Table 3) (HATZFELD, 1978; GRIMISON and CHENG 1986).

Table 3

Hypocentral coordinates of intermediate shocks before 1976 with fault-plane solutions

Date	Lat.	Long.	Depth	Magn. (km)	Ref.
02–07–1972	36.0N	4.6W	100	4.0	1
13–06–1974	36.9N	4.1W	69	4.2	1
07–08–1975	36.4N	4.4W	94	5.1	1, 2

References:
1. HATZFELD, 1978
2. GRIMISON and CHENG, 1986

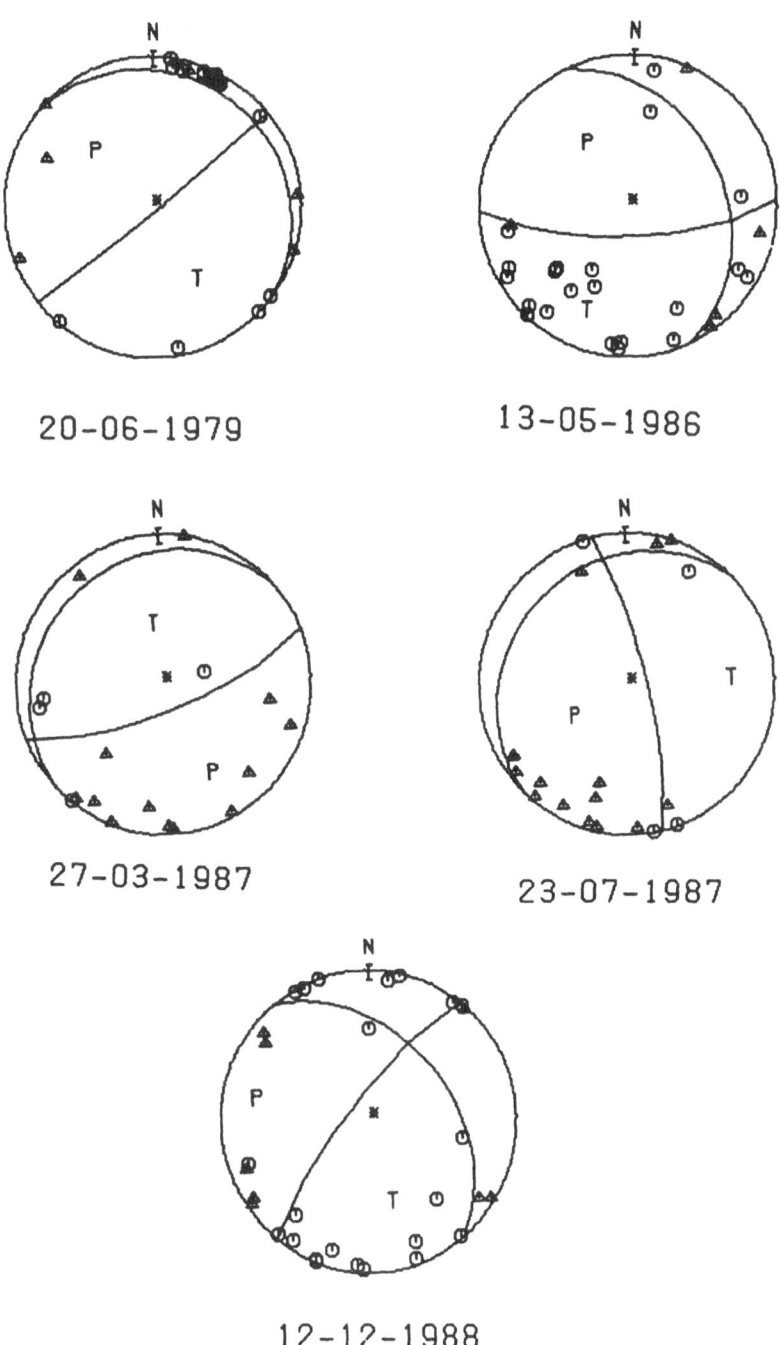

Figure 4

Fault-plane solutions of the intermediate depth shocks. Stereographic projection of the lower hemisphere are represented. Triangles correspond to dilatations and squares to compressions.

The solution obtained for shock 07–08–1975 is similar to our solution for event 7, but the plane obtained by HATZFELD (1978) and GRIMISON and CHENG (1986) in direction NW-SE is nearly vertical and the *P* axis is horizontal. The fault-plane solution for the earthquake of 13–06–1974 shows a vertical N-S plane and another dipping about 45° in E-W direction, while the pressure axis is oriented in SE direction with small dip. The solution for the event of 02–07–1972 is not well determined, due to only 9 observations. The problem for the determinations of fault-plane solutions for earthquakes of magnitude about 4, at that time, is due to the poor coverage of stations. Sometimes, data have been taken from ISC (International Seismological Center, Edinburgh) or stations bulletins and may contain errors.

Seismic Moment and Dimensions

Seismic moment and source dimensions were estimated from digital records of the Spanish National Seismic Network. The vertical component of the 25 seismic stations of this network are currently digitized at 100 samples per second and stored on magnetic tapes. These digital records were windowed, detrended, Fourier transformed and corrected for instrumental response. Amplitude spectra of *P*-waves were then calculated for all those stations that recorded the earthquakes on-scale. Unfortunately the present 16-bit analog-digital converters used on the network have a rather poor dynamic range and for any earthquake only a subset of stations provide useful unsaturated seismic records. 6 s windows around the *P*-wave arrival were hand-picked in order to calculate amplitude spectra. Examples of typical spectra obtained at three stations of the network, for two intermediate-depth earthquakes, are shown in Figure 6.

The seismic moments were calculated from the usual formulas relating low frequency trend of the spectra to the seismic moment. For lack of a more detailed model, we assume that waves were generated and propagated in a homogeneous upper mantle model with density $3.4 \, \text{g.cm}^{-3}$ and *P*-wave velocity $8.1 \, \text{km s}^{-1}$. Hypocentral distances were calculated from depths determined by I.G.N. These magnitudes are not really calibrated, because they were calculated using parameters appropriate for shallow focus earthquakes. The seismic moments for all events were very similar in spite of rather significant variations in magnitude. The magnitude of shock number 5 was probably underestimated because the amplitudes observed at stations with equal hypocentral distances were similar for this event and for the other two. The precision in estimation of seismic moments is of the order of a factor of two.

Source dimensions were estimated for the four intermediate events using Brune's circular source model (BRUNE, 1970). Corner frequencies were estimated by direct measurement of the intersection of the low and high frequency trends of the spectra plotted in double logarithmic coordinates. It should be noted that at some stations

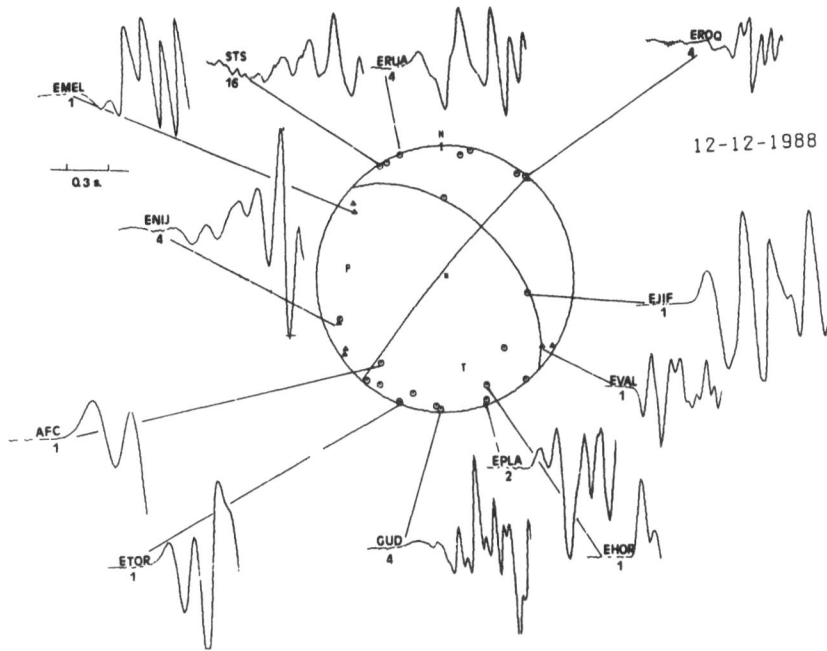

12-12-1988

Figure 5

Fault-plane solution and *P*-wave form for event 12 December, 1988. The records correspond to Spanish digital stations. Numbers show the amplification factors for the amplitude scale.

we found some indication of intermediate frequency trends that were ignored. In all cases the corner frequency was defined by the steepest high-frequency slope. Converting the corner frequency into seismic source radii is delicate, not only coefficients used in different models vary over factors larger than two; but in the present case, there is the further problem of adopting a reference value for the *P*-wave velocity at the hypocenter. Table 4 shows the source dimensions we estimated by the previous method. Again all events appear to be similar with little variation in radius and seismic moment. The values obtained for the seismic moment and dimensions agree with those calculated by other authors for earthquakes of similar magnitude (THATCHER and HANKS, 1973; ARCHULETA *et al.*, 1982).

Table 4

Seismic moments and source dimensions of Spanish intermediate and deep earthquakes

Event	M	h(km)	M_0(Nm)	2r(km)	N
13–05–1986	4.5	90	$(2.2 \pm 1.4) \times 10^{15}$	1.0	3
27–03–1987	3.5	69	$(2.4 \pm 1.3) \times 10^{15}$	1.2	4
23–07–1987	3.9	100	$(2.1 \pm 1.0) \times 10^{15}$	--	6
12–12–1988	4.2	70	$(2.0 \pm 1.2) \times 10^{15}$	0.8	4
08–03–1990	4.8	640	$(1.2 \pm 0.7) \times 10^{16}$	2.6	10

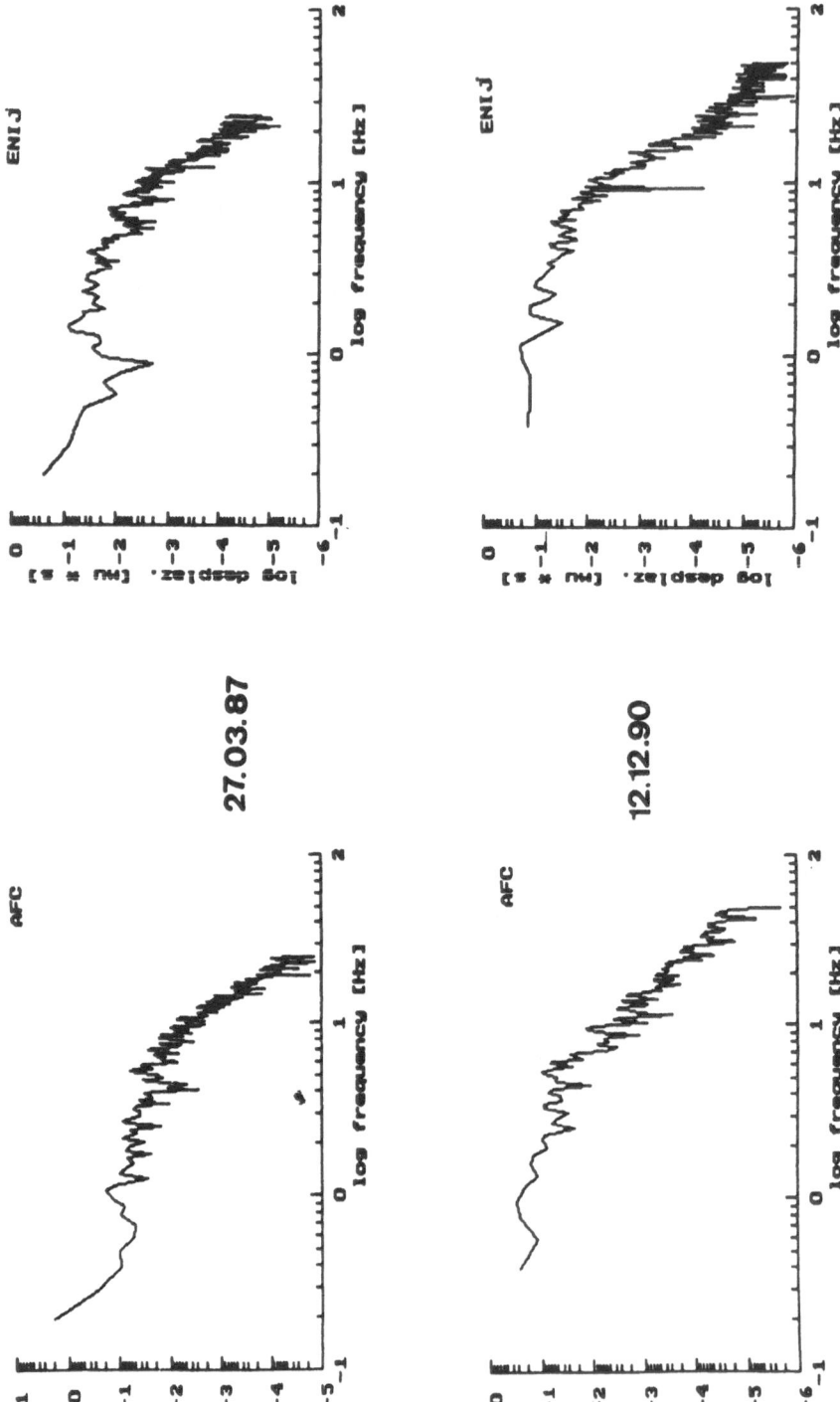

Figure 6

Amplitude spectra of intermediate depth shocks estimated from digital records of the Spanish Seismological Network, I.G.N.

Deep Earthquakes

The occurrence of deep shocks, at a depth of about 650 km, below southern Spain, near Granada is well known. The first recorded earthquake took place on 29 March, 1954, with magnitude 7. In the same place, shocks with lower magnitudes occurred on 30 January, 1973, and 8 March, 1990. In Table 1 and Figure 11, the hypocentral coordinates and the locations are shown. The deep earthquake of 1954 has been the subject of several studies by HODGSON and COCK (1956), BONELLI and ESTEBAN-CARRASCO (1957), CHUNG and KANAMORI (1976), UDÍAS et al. (1976) and GRIMISON and CHEN (1986).

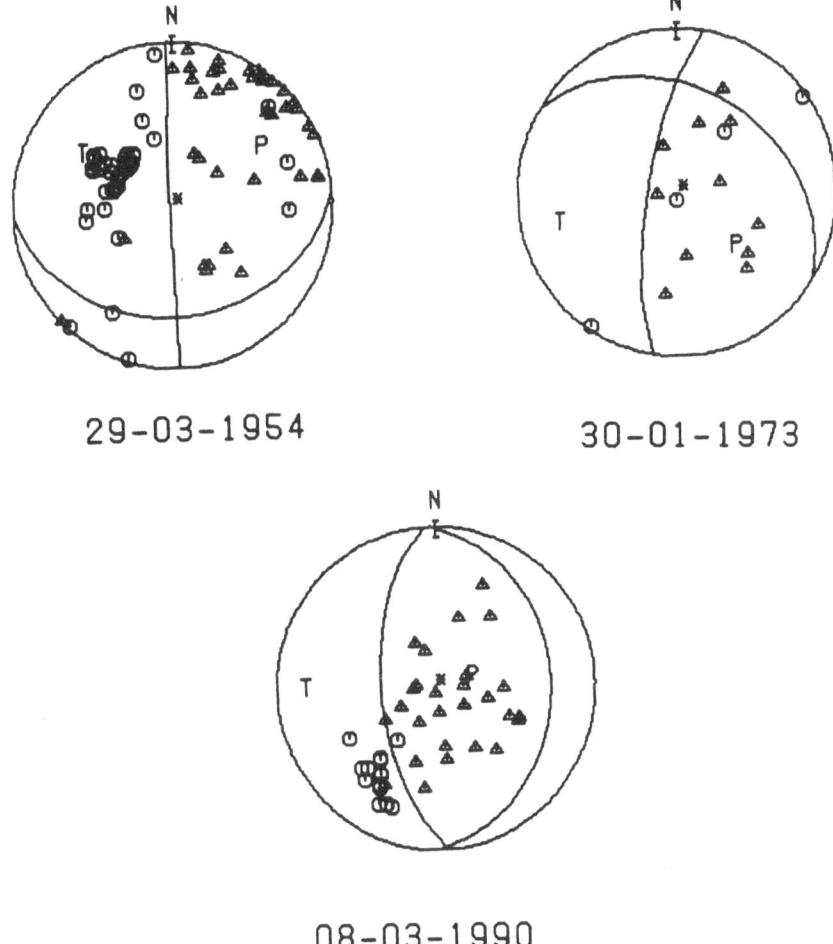

29-03-1954   30-01-1973

08-03-1990

Figure 7

Fault-plane solutions for deep shocks. Stereographic projection of the lower hemisphere are represented. Triangles correspond to dilatations and squares to compressions.

Fault-plane Solutions

In Figure 7 and Table 2, the fault-plane solutions of the three deep earthquakes are shown. For the 1954 and 1973 events they have been redetermined. The data used for the 1954 event are the polarities of *P* waves taken from HODGSON and COCK (1956) from teleseismic distances. For the 1973 and 1990 events, first motion data were read from the original records of stations at regional distances.

The three solutions present a fault plane in N-S direction, vertical for the 1954 event and dipping 74° and 62° west for the 1973 and 1990 events. The other plane is poorly defined with similar dip for the three events, about 30°, and E-W orientation for 1954, NW-SE for 1973 and N-S for 1990. The pressure axis dips to the east and the tension axis to the west. The solutions obtained by HODGSON and COCK (1956) and CHUNG and KANAMORI (1976) for the 1954 shock agree with our solution, but their E-W plane is more horizontal. The fault-plane orientation for the 1973 earthquake agrees with that obtained by UDÍAS *et al.* (1976) and differs with that of GRIMISON and CHEN (1986). Our solution has a more vertical N-S plane, the second plane has more N-S orientation.

Figure 8 shows the fault-plane solution and *P*-wave forms which form digital records of the Spanish stations of the 8 March, 1990 earthquake. The horizontal scale (time) is the same for all stations but the amplitudes are multiplied by

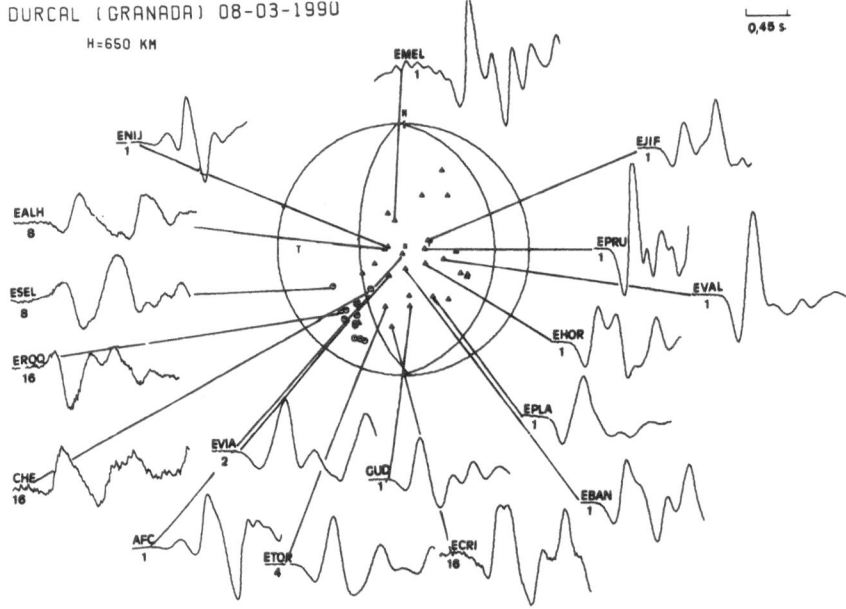

Figure 8
Fault-plane solution and *P*-wave forms for event 8 March, 1990. The records correspond to Spanish digital stations and numbers to the amplification factor for the amplitude scale.

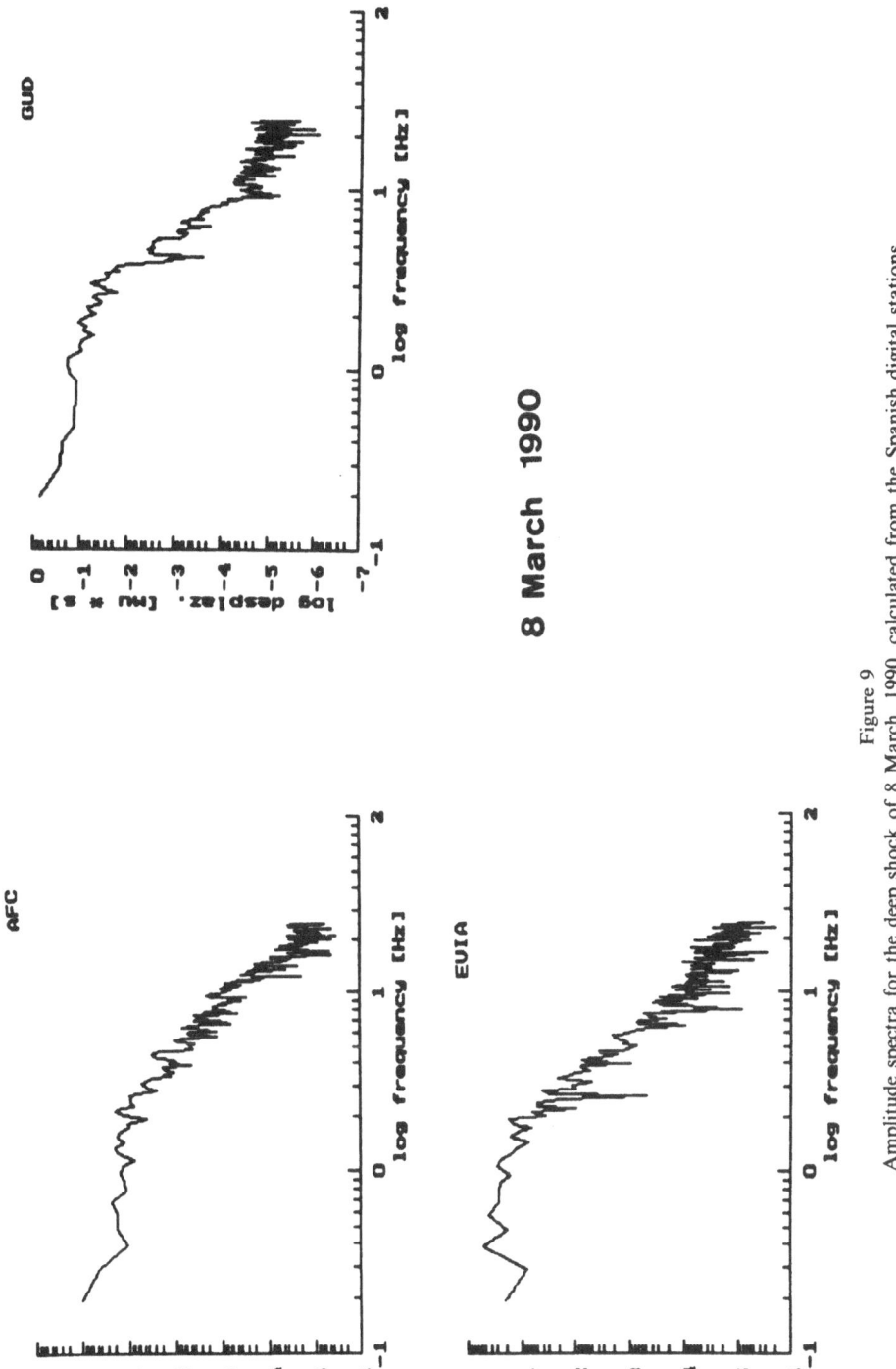

Figure 9

Amplitude spectra for the deep shock of 8 March, 1990, calculated from the Spanish digital stations.

different factors indicated in the figure. For some stations, the wave forms are similar, for example: EVIA, GUD and ETOR; ENIJ and AFC; EPLA and EVAL. In general, the *P* waves have a simple form with a single pulse followed after about 0.5 to 0.8 seconds by a second arrival. According to CHUNG and KANAMORI (1976), the wave forms for the 1954 earthquake are very complex and the earthquake was considered a multiple event. A comparison between the *P*-wave forms of the intermediate depth shock (Figure 5) and those of the deep one (Figure 8) shows that the latter is of simpler form. This may be due to the fault that for the deep earthquake, waves are less affected by heterogeneities in the crust and upper mantle.

Seismic Moment and Dimensions

Seismic moment and some source dimensions for the deep event were estimated using the same technique employed for the intermediate depth events. The seismic moment estimate is $(1.7 \pm 0.7) \times 10^{16}$ Nm and the dimensions of 2.6 km.

The 1973 and 1990 events have been recorded on the WWSSN short-period instrument of TOL (Toledo) station. Records for the *S* waves are shown in Figure 10 and they are similar. Spectra have been calculated, resulting in similar values for the amplitudes of the flat part. Both events must, then, have similar seismic moments. The magnitude given for the 1973 event (4.0) must be corrected to a value similar to that of the 1990 event (4.8).

Seismotectonic Interpretation

Figure 11 shows the fault-plane solutions for intermediate and deep shocks in southern Spain and the most important geological faults of this area. The solutions for intermediate shocks in southern Spain and northern Morocco are consistent with vertical fault planes oriented in E-W direction (event 4), NE-SW direction (3, 5 and 7) and N-S (6) and pressure axes dipping about 45° in NW direction (3, 4, 7) in southern Spain and SW direction in northern Morocco (6). The orientation of the *P* axis for shock 5 does not agree with this general western trend and is rotated 180° facing east. For the deep shocks the fault-plane solutions have a nearly vertical plane very well defined with strike in NS direction. The pressure axis dips 45° to the east.

Pressure axes projected on a vertical plane are represented in Figure 12 for intermediate (Fig. 12a) and deep shocks (Fig. 12b). For the intermediate shocks the vertical plane is oriented in NW-SE direction with center at the Málaga coast. Earthquake No. 6 is not represented because of the long distance to the plane of projection. Three of the pressure axes dip about 45° to NW. The only exception is that of event No. 5 with its pressure axis dipping to the SE. For the deep shocks, the vertical plane is oriented in E-W direction and the pressure axes dip about 45°

TOL SP E-W 30 ENERO 1973

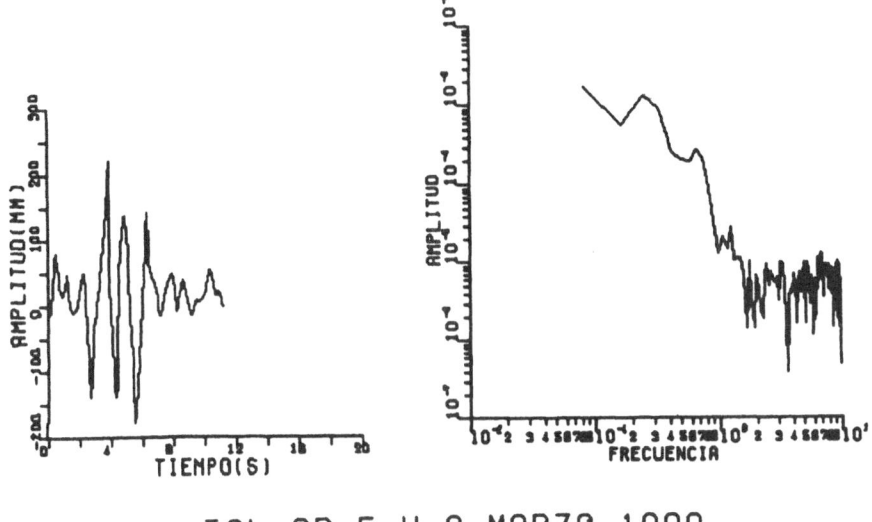

TOL SP E-W 8 MARZO 1990

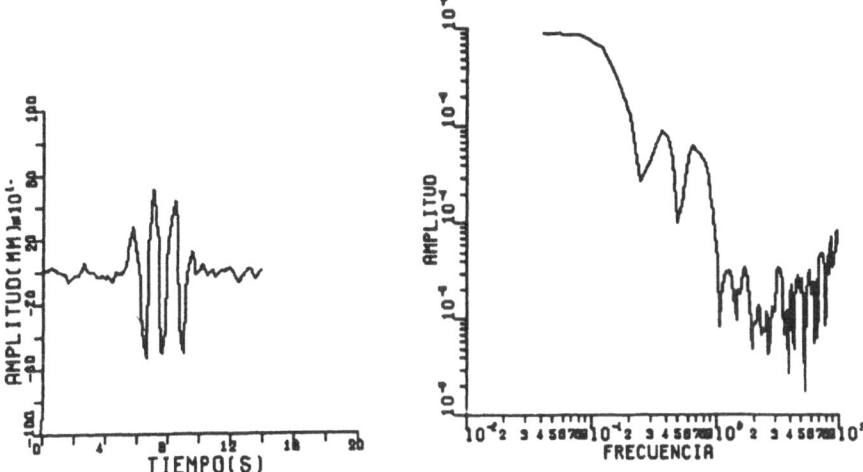

Figure 10
Amplitude spectra for the deep shocks of 30 January, 1973 and 8 March, 1990 estimated for the short period WWSSN records in Toledo station.

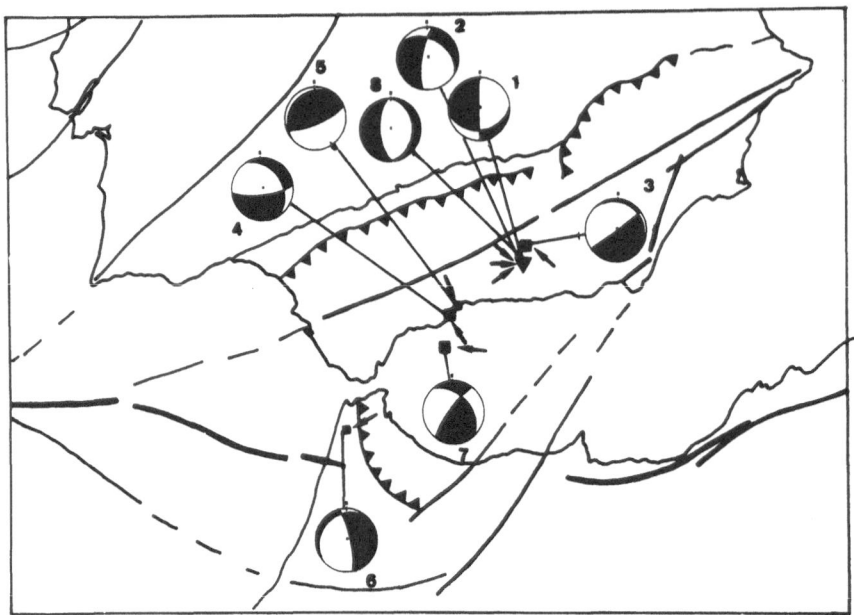

Figure 11
Seismotectonic framework for this region. Fault-plane solutions correspond to intermediate and deep
shocks. Arrows show the pressure axis orientation.

to the east. This direction is opposite that of the *P* axes of intermediate shocks, which dip to the W and NW. From the point of view of their focal mechanisms, then, these two sets of earthquakes seem to have a different origin.

In a recent paper, BUFORN *et al.* (1988) showed that focal mechanism of shallow shocks in southern Spain may be interpreted in terms of regional compressive stresses in NNW-SSE direction. This compression is produced by the collision between the plates of Eurasia and Africa which produces a deformation zone in southern Spain, northern Morocco and Algeria. If intermediate depth earthquakes are caused by this collision one would expect an E-W trend. However, the trend observed is almost N-S from Granada to northern Morocco (Figure 2). Focal mechanisms of the intermediate earthquakes ($30 < h < 150$ km) indicate a compression dipping 45° in NW to W direction. This situation may be interpreted in terms of an almost vertical slab of brittle and cold material extending from the surface to about 150 km (Figure 3) subducted or sunk from the east or southeast. Horizontal movements of crustal material toward the west are supposed to have been of great importance in the formation of the Beltic and Rif cordilleras and the thinning of the crust in the Alboran Sea (SANZ DE GALDEANO, 1990). These processes could also be the cause of forcing the material in front of the advancing units to sink into the mantle in the N-S front. The maximum depth of intermediate shocks, about 150 km, would indicate the vertical extension of the subducted slab. If relative plate

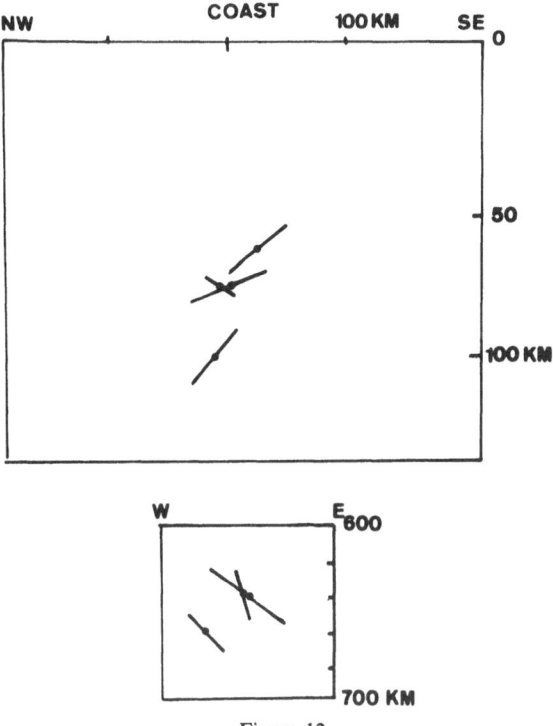

Figure 12
Vertical cross section of the pressure axes of intermediate and deep shocks.

motion in this region is approximately 1 cm/yr, the subducted material down to 150 km would have been formed in about 15 million years. This will represent, then, the result of relatively recent processes.

Between 150 km and 600 km no seismic activity has been detected. At 640 km depth a nest of seismic foci has been located in a very punctual zone beneath Granada and seems to be independent of the shocks which occur in the first 150 km. It has been shown that their focal mechanisms are different with the P axes pointing to the east. Their origin must be related to a detached block of lithospheric material still sufficiently cold and rigid to generate earthquakes (UDÍAS et al., 1976; CHUNG and KANAMORI 1976; GRIMISON and CHEN, 1986). Assuming that for deep earthquakes pressure axes point down dip the subducted slab, the orientation of the pressure axis for these earthquakes dipping east, makes it difficult to relate their orientation to the actual plate collision which in this region is N-S to NW-SE. The process of sinking through mantle material, however, may explain by itself changes in the orientation of stress axes in the lithospheric block. The present mechanism of these earthquakes may not be related, then, to the stresses in the subducted lithosphere at the time of its formation near the surface. This original subduction process must have taken place before the start of that one responsible for the intermediate depth shocks, because of the gap existing between them. The slab of material would have, then, been detached from the surface and sunk to its present depth. At present only a very small volume seems to maintain sufficient strength to generate earthquakes.

Intermediate and deep seismicities are located at a region where an anomalous distribution of seismic velocity has been detected by surface wave analysis (PANZA *et al.*, 1980) and teleseismic *P* wave tomography (SPACKMAN, 1986). Recently, a tomographic study (BLANCO and SPACKMAN, 1991) has shown the existence of an anomalous high velocity region below southern Spain extending from 200 to 700 km in depth. The presence of this anomalous body in the upper mantle at this region must be related to the occurrence of intermediate and deep seismic activity. The detailed structure of this anomalous region is not sufficiently known. The presence of intermediate shocks in the area outside the region studied here (North Morocco and Gulf of Cádiz) complicates the problem. Models of geological evolution of the Iberian peninsula and the Alboran Sea must, however, take into account the presence of this deep anomalous structure and the seismic activity at intermediate and very deep depths associated with it.

Acknowledgements

The authors wish to thank Dr. J. Mezcua and J. Rueda, Instituto Geográfico Nacional Madrid, for providing the seismicity records and maps from the I.G.N. Seismicity Data File and J. M. Tejedor who provided the software necessary to read the Spanish digital station records. This work has been supported in part by the Direccion General de Investigacion Científica y Tecnológica, project PB-89-0097 and by the European Economic Community, project SC1*0176-C (SMA). Publication 329, Department of Geophysics, Universidad Complutense de Madrid.

REFERENCES

ARCHULETA, R., CRANSWICK, E., MUELLER, C., and SPUDICH, P. (1982), *Source Parameters of the 1980 Mammoth Lakes, California, Earthquake Sequence*, J. Geophys. Res. *87*, 4595–4607.
BLANCO, M. J., and SPACKMAN, W. (1991), *The P Velocity Structure of the Mantle below the Iberian Peninsula: Evidence for Subducted Lithosphere below Southern Spain*, Tectonophys. (in press).
BONELLI, J. M., and ESTEBAN CARRASCO, L. (1957), *El sismo de foco profundo de 29 de marzo de 1954 en la falla de Motril*, Instituto Geográfico Nacional, Madrid, 28 pp.
BRUNE, J. N. (1970), *Tectonic Stress and the Spectra of Seismic Shear Waves from Earthquakes*, J. Geophys. Res. *75*, 4997–5009.
BUFORN, E., UDÍAS A., and MEZCUA, J. (1988), *Seismicity and Focal Mechanisms in South Spain*, Bull. Seismol. Soc. Am. *78*, 2008–2024.
CHUNG, W., and KANAMORI, H. (1976), *Source Process and Tectonic Implications of the Spanish Deep-focus Earthquake of 29 March 1954*, Phys. Earth Plan. Int. *13*, 85–96.
GRIMISON, N., and CHENG, W. (1986), *The Azores-Gibraltar Plate Boundary: Focal Mechanisms, Depths of Earthquakes and their Tectonic Implications*, J. Geophys. Res. *91*, 2029–2047.

HATZFELD, D. (1978), *Etude sismotectonique de la zone de collision Ibero-Maghrebine*. Doctoral Thesis, Université de Montpellier, 282 pp.

HODGSON, J.H., and COCK, J. I. (1956), *Direction of Faulting in the Deep Focus Spanish Earthquake of March 29, 1954*, Tellus *8*, 321–328.

MCKENZIE, D. (1972), *Active Tectonics of the Mediterranean Region*, Geophys. J. R. Astr. Soc. *30*, 109–185.

MCKENZIE, D. (1978), *Active Tectonics of the Alpine-Himalayan Belt: The Aegean Sea and Surrounding Regions*, Geophys. J. R. Astr. Soc. *55*, 217–254.

MEZCUA J., and UDIAS, A. (1991), *Seismicity, Seismotectonics and Seismic Risk of the Ibero-Maghrebian Region*, Instituto Geográfico Nacional, Madrid, 390 pp.

MUNUERA, J. M. (1963), *Datos básicos para un estudio de sismicidad en la región de la Peninsula Ibérica*, Mem. Inst. Geog. Cat., Madrid 32, 93 pp.

PANZA, G. F., MUELLER, S., and CALGAGNILE, G. (1980), *The Gross Features of the Lithosphere-astenosphere System in Europe from Seismic Waves*, Pure Appl. Geophys. *118*, 1209–1213.

PAPAZACHOS, B. (1973), *Distribution of Seismic Foci in the Mediterranean and Surrounding Area and its Tectonic Implication*, Geophys. J. R. Astr. Soc. *33*, 421–430.

PAPAZACHOS, B., and COMNINAKIS, P. (1977), *Modes of Lithospheric Interaction in the Aegean Area. Structural History of the Mediterranean Basins*, Split (Yugoslavia) Technip. Paris, 319–332.

RITSEMA, A. R. (1969), *Seismic Data of the West Mediterranean and the Problem of Oceanization*, Verhand. K. Ned. Geol. Minjnbown Gen. *26*, 105–120.

RITSEMA A. R. (1972), *Deep Earthquakes of the Thyrhenian Sea*, Geologie in Mijnbouw 51 (5), 541–545.

SANZ DE GALDEANO, C. (1990), *Geologic Evolution of the Betic Cordillera in the Western Mediterranean, Miocene to the Present*, Tectonophys. *172*, 107–119.

SPACKMAN, W. (1986), *Subduction beneath Eurasia in Connection with the Mesozoic*, Tethys. Geologie en Mijbonw *65*, 145–153.

THATCHER, W., and HANKS, T. (1973), *Source Parameters of Southern California Earthquakes*, J. Geophys. Res. *78*, 8547–8576.

UDÍAS, A., LOPEZ ARROYO, A., and MEZCUA, J. (1976), *Seismotectonics of the Azores-Alboran Region*, Tectonophys. *31*, 259–289.

UDÍAS, A. (1982), *Seismicity and Seismotectonic Stress Field in the Alpine-Mediterranean Region*, Alpine Mediterranean Geodynamics, Geodynamics Series *7*, 75–82.

UDÍAS, A., BUFORN, E., and RUIZ DE GAUNA, J. (1989), *Catalogue of Focal Mechanisms of European Earthquakes*, Dept. of Geophysics, Universidad Complutense, Madrid, 274 pp.

(Received November 10, 1990, revised/accepted February 20, 1991)

PAGEOPH, Vol. 136, No. 4 (1991)

0033–4553/91/040395–10$1.50 + 0.20/0
© 1991 Birkhäuser Verlag, Basel

Spanish National Strong Motion Network. Recording of the Huelva Earthquake of 20 December, 1989

E. Carreño,[1] J. Rueda,[2] C. López Casado,[2] J. Galán[1] and J. A. Peláez[2]

Abstract — In recent years a network of 30 accelerographs has been installed through the zones of highest seismic activity of Spain. For the first time, digital strong motion records have been obtained in Spain, with a maximum horizontal acceleration value of 0.06 g. A comprehensive study is made of the strong motion recordings of an earthquake which occurred in southwest Spain, on December 20, 1989. The isoseismal map is drawn and the data confirm the main attenuation directions in the area observed in other shocks.

Key words: Strong-motion, focal mechanism, attenuation, acceleration–intensity, spectral response.

Introduction

The INSTITUTO GEOGRÁFICO NACIONAL of Spain, (I.G.N.), began in the last decade to establish a network of strong motion instruments distributed through regions of high seismic activity. Initially, the instruments were 9 with analogue recording, SMA-2 Kinemetrics. The program was interrupted for 3 years, and a new one was started as digital accelerographs became available on the market. The installation of the digital equipment began in 1989 with 11 Spanish instruments ACD-3 Ofiteco. Later, we continued with the installation of 10 digital instruments SSA-1, Kinemetrics.

The first digital accelerogram in Spain was obtained as a consequence of the December 20, 1989, earthquake of 4.8 magnitude that took place in South Spain at the boundary with Portugal. As an area of seismotectonic interest and because a new digital accelerograph was in operation near the epicenter, a comprehensive study was carried out.

The focal mechanism of this event is analyzed and an interpretation based on the tectonics of the region is given. In order to compare the intensity values with other parameters a detailed macroseismic survey was carried out also showing characteristics related to the general tectonics of the region.

---

[1] Instituto Geográfico Nacional, General Ibañez Ibero, 3, 28003 Madrid, Spain.
[2] Facultad de Ciencias, Universidad de Granada, Fuentenueva s/n, 18071 Granada, Spain.

Figure 1
Spanish National Strong Motion Network. Triangles represent digital instruments while squares are
analogical.

Network Instrumentation

Figure 1 shows the actual distribution of strong motion instruments in Spain. The analogical accelerographs are the triaxial model SMA-2 of Kinemetrics. These instruments may have their acceleration vertical component adjusted to a trigger level. The natural frequency of these instruments is roundly 30 Hz with a 1 g full scale.

The digital instruments are of two types: ACD-3 (Ofiteco) and SSA-1 (Kinemetrics); both types have a large dynamic range and solid state memory, allowing the recording of a complete accelerogram.

For the establishment of the instruments, criteria have been followed, based on the maximum intensity maps, isolation of buildings, where possible in order to obtain free field recordings. Also, for future implementation, the possibility of connecting by modem with the seismic center in Madrid has been considered.

20 December, 1989 Earthquake

Focal Parameters

The focal parameters of the 20 December, 1989, earthquake are listed in Table 1. The epicenter is located approximately at Isla Cristina (Huelva), at a hypocentral depth of about 23 km.

After the earthquake, questionnaires were sent to an extensive area of the South and South-West of the Peninsula; knowing that it had been felt at distant sites from the epicenter. The isoseismal map (Figure 2) has been drawn from 991 completed

Table 1

Focal parameters of the 20 December, 1989 earthquake.

Orig. Time	4h 15 m 05.99 s
Latitude	37°17'.2 N
Longitude	7°19'.7 W
Depth	23 km
Magnitude (Mb)	4.8
Max. Intensity	VI
N° Observations	25

questionnaires. From the isoseismal map, we have computed the corresponding areas for the zones of each intensity value. Since the epicenter is near the sea, as can be seen on the isoseismal map, the isoseismal lines do not close southwards. For this reason we have had to make an approximation to compute this area. From the isoseismal map we can observe a lower attenuation in the higher intensities along the Guadiana river in a north-south direction.

From the intensity map, we have obtained the average radius of the circular area of each grade of intensity, and have drawn the attenuation curve shown in Figure 3 together with attenuation curves obtained by MARTIN (1984) for South Spain from near earthquakes and from distant earthquakes from the Azores-Gibraltar fault. This figure shows that the attenuation for this earthquake is lower than the two cases studied by MARTIN (1984). One possible explanation, which has been observed in several earthquakes, is that there are preferred directions which coincide with the maximum of the nonconsolidated sediments.

Seismicity of the Area

This earthquake is located near the northern boundary of the active region of South Spain (Figure 4). The main feature of this region is that it constitutes part of the contact between Africa and Eurasia plates with moderate earthquake magnitude ($M < 5$), but also the occurrence of large earthquakes separated by long time intervals with maximum intensities of *IX* and *X*. The distribution of epicenters for the period 1965 to 1985 shows that earthquakes are located south of the Cádiz-Alicante fault and that they can be associated with observed geological faults (BUFORN *et al.*, 1988).

The mechanism of this earthquake (I.G.N., 1991) shows a strike-slip mechanism; with a horizontal pressure axis in the NW-SE direction. The fault plane with a strike in the E-W direction has been chosen in accordance with the dominant geological lineaments in the region. Motion on that plane is right lateral with the north block moving east in agreement with the horizontal motion in the Azores-Gibraltar fault.

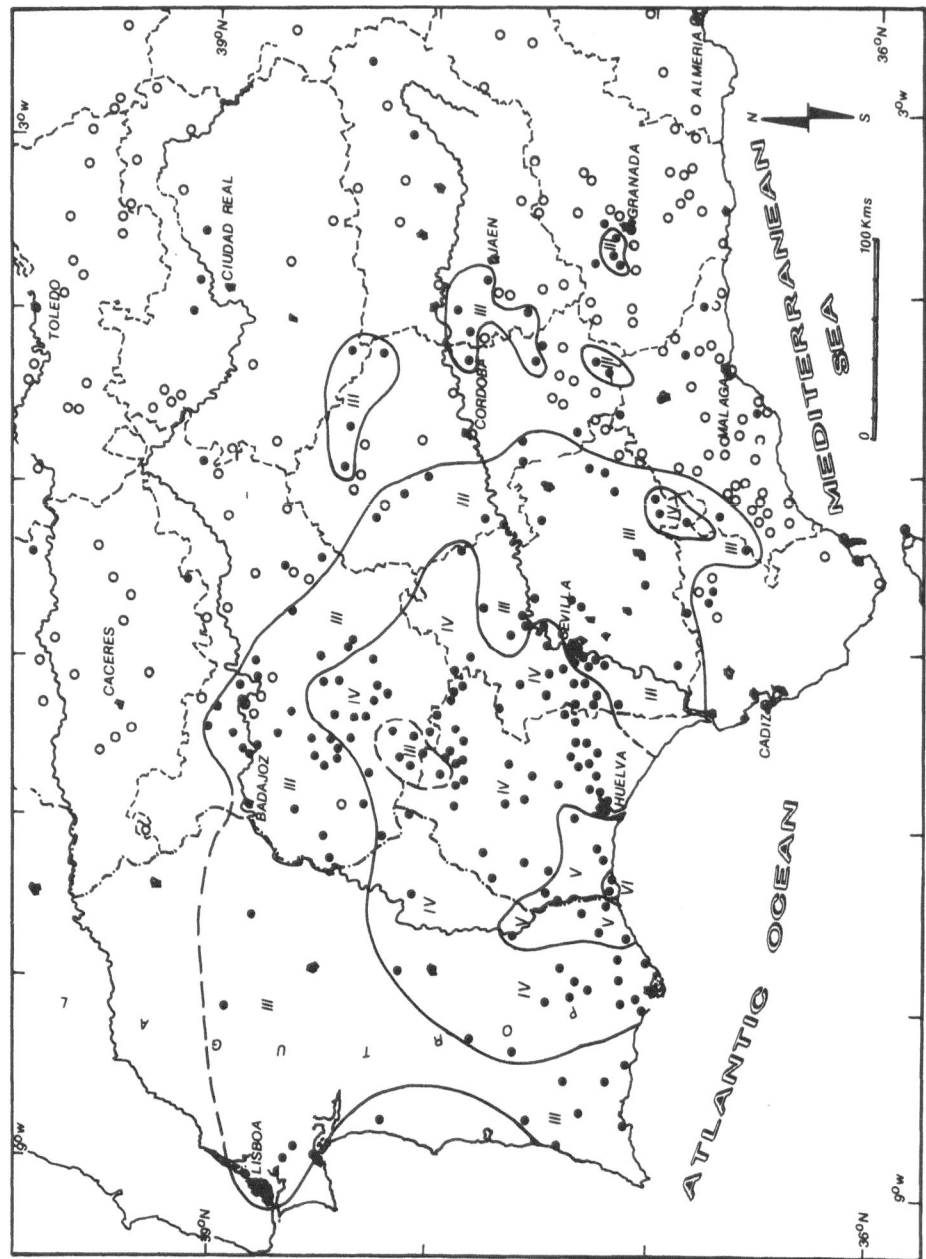

Figure 2

Isoseismal map of 1989 December 20, earthquake. Open circles show sites where the earthquake was not felt.

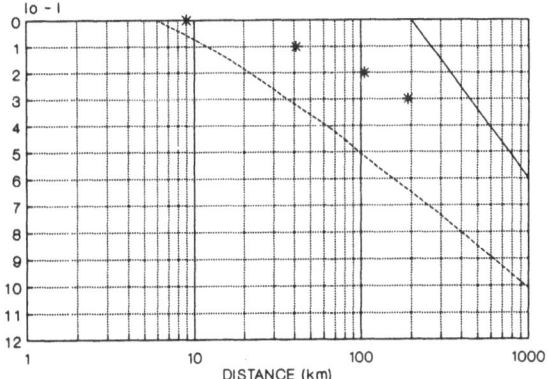

Figure 3
Attenuation values of 20 December, 1989 earthquake (asterisk) compared with those obtained taking the influence of Azores-Gibraltar zone (dotted) and without (continuous) as given by MARTIN (1984).

Comparing it with the mechanism of the 10–20–1986, and 3–11–1987 earthquakes, the orientation of P axes are in agreement with the general idea of a general regional stress with NW-SE compression in the area (Figure 4). In contrast, the 5–26–1985 earthquake corresponds to a reverse faulting with planes oriented North-South, and the P axis in E-W direction.

Analysis of Strong Motion Data

The strong motion instrument located in Cartaya (Huelva), was the one nearest the epicenter (19 km). The accelerograph, an Ofiteco ACD-3P digital instrument, was the only one that triggered. The accelerograph is placed on a concrete peer, located in the basement of a building. Geologically, the soil is composed of recent marine terrace of Pliocene sandy loam.

Only the two horizontal components were processed because the vertical sensor failed.

Direct numerical integration of the accelerations appears to be contaminated by long periods. We have obtained the acceleration, velocity and displacement using parabolic correction and a band-pass filter between 0.7 and 30 Hz.

Figure 5, shows the acceleration, velocity and displacement of the EW and NS components respectively after having been filtered between 0.7 and 30 Hz. The maximum values are listed in Table 2 as well as values from analogical accelerograms of the 24 June, 1984 earthquake, magnitude 5.0, CARRENO et al. (1988).

In Figure 6, we observe the trilogarithmic plots of pseudovelocity for 0, 5 and 10% damping; corresponding to the E-W and N-S components.

The maximum values encountered in these accelerograms are the largest recorded in Spain and they allow, with the previous analogue records in Granada

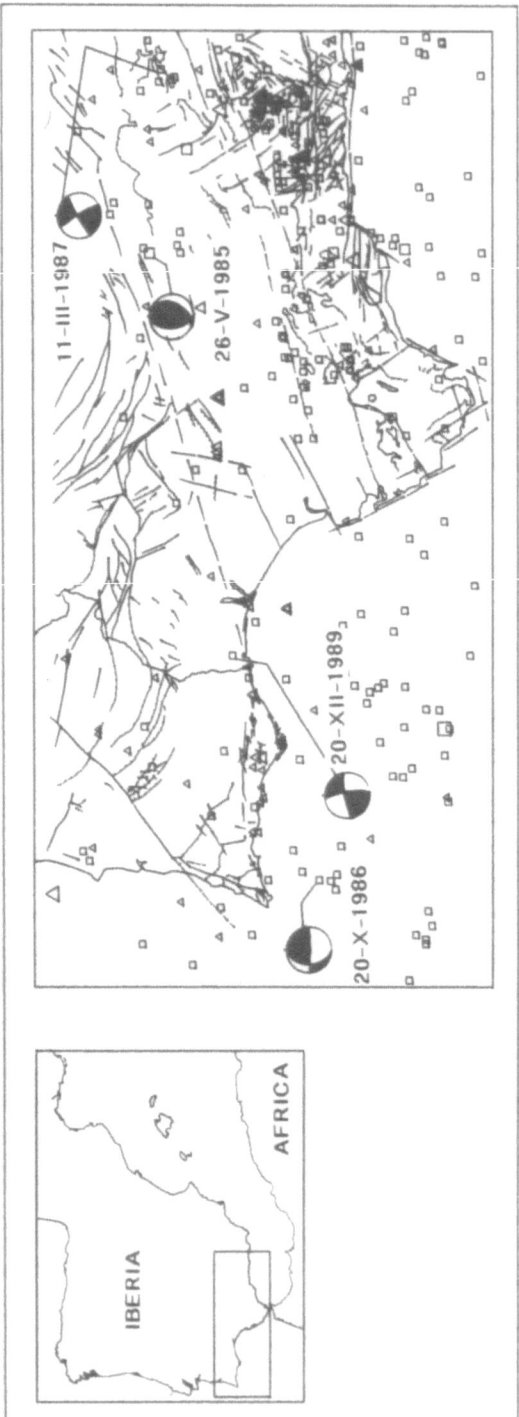

Figure 4

Seismicity map from Seismic Data Bank (I.G.N.), with main fractures for the zone (SANZDE GALDEANO, 1983). The mechanism corresponds to 11–20–89, 10–20–86, 3–11–87, and 5–26–85 earthquakes (BUFORN *et al.*, 1988).

ACCELERATION WITH PARABOLIC CORRECTION. COMPONENT EW
BUTTERWORTH BAND PASS FILTER CONSTANTS. F(LOW) 0.7 F(HIGH) 30.0 IT. 6

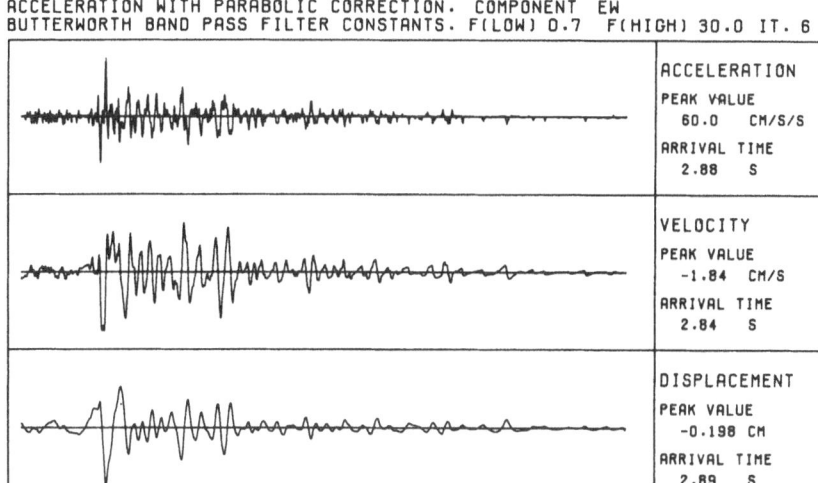

TIME (S)

ACCELERATION WITH PARABOLIC CORRECTION. COMPONENT NS
BUTTERWORTH BAND PASS FILTER CONSTANTS. F(LOW) 0.7 F(HIGH) 30.0 IT. 6

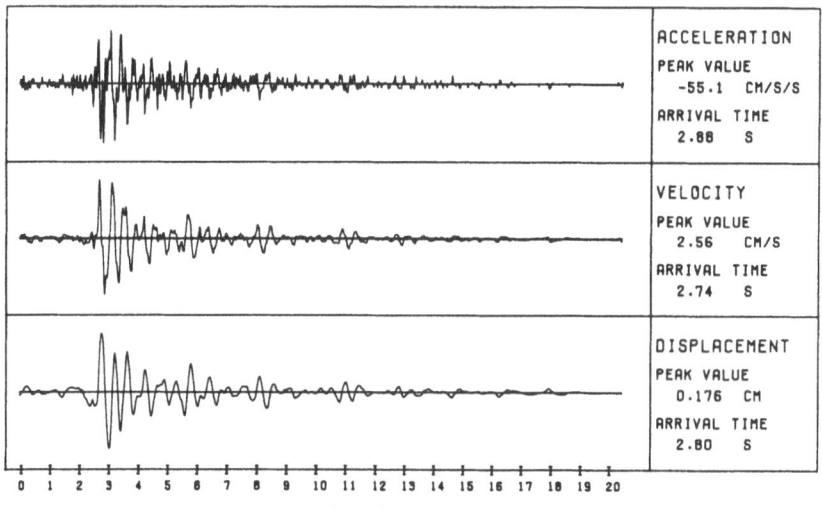

TIME (S)

Figure 5
Acceleration, velocity and displacement of the EW and NS components. A parabolic correction and a
Butterworth filter band pass of order ten, between 0.7 and 30 Hz have been made.

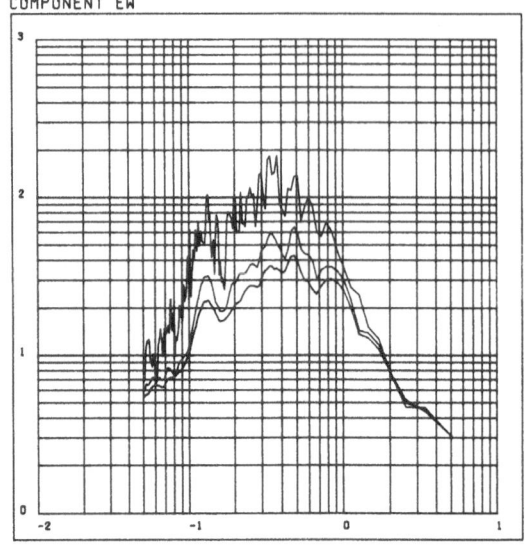

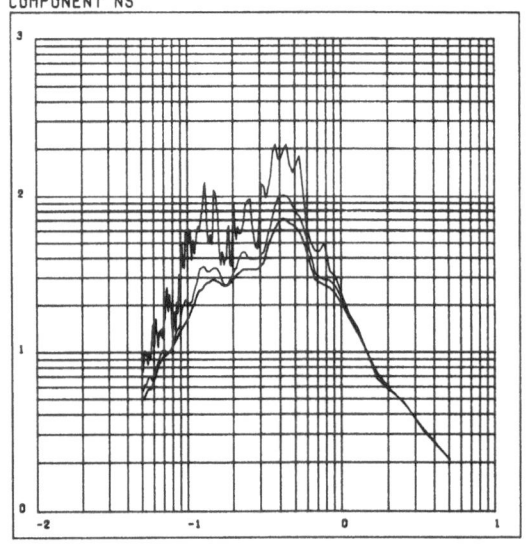

Figure 6
Spectral response for the EW and NS components and damping curves of 0, 5 and 10%.

for the 1984 earthquake, the accomplishment of some correlations for earthquake engineering in which acceleration is one parameter. In Figure 7 the values of maximum acceleration for the two earthquakes (20–12–89 and 24–6–84) are plotted against intensity. In the same figure several relations by different authors, COULTER *et al.* (1973), AMBRASEYS (1975) and TRIFUNAC and BRADY (1975), are given.

Table 2

Acceleration, velocity and displacement for 24 June, 1984 and 20 December, 1989.

Earthquake	Stat.	Comp.	cm/s²	cm/s	cm	Ep.D. (k)	Mg.
24–6–84	Alhama	E–W	12.9	0.31	0.02	28.4	5.0
		N–S	15.3	0.19	0.01		
		Z	35.5	0.70	0.09		
24–6–84	Beznar	E–W	19.8	0.77	0.06	20.0	
		N–S	19.7	0.78	0.04		
		Z	29.0	1.34	0.09		
24–6–84	Santa Fe	E–W	35.0	1.59	0.09	38.6	
		N–S	29.7	1.72	0.07		
		Z	23.4	0.56	0.03		
20–12–89	Cartaya	E–W	60.0	1.85	0.19	17	4.8
		N–S	55.0	2.56	0.17		

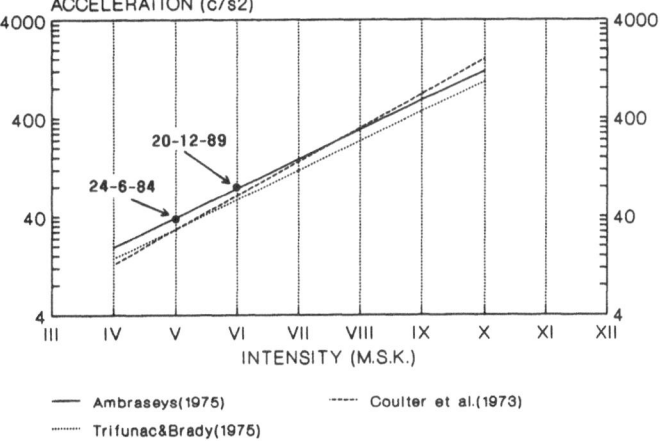

Figure 7

Maximum intensity, horizontal acceleration relation following several authors. It shows the corresponding values for the earthquakes of 20 December, 1989, and 24 June, 1984.

The best agreement in the intermediate range of intensity values is with the curve given by AMBRASEYS (1975). Beyond intensity VI, the tendency could be fitted also by COULTER *et al.* (1973).

Conclusions

From the isoseismal map we can observe a slow attenuation in the higher intensities along the Guadiana river in a north-south direction.

This fact can also be appreciated in the isoseismal map obtained for the following earthquakes: 15 March, 1964, 28 February, 1969, 23 May, 1980 and 7 ctober, 1983 (MEZCUA, 1982).

The strike-slip mechanism confirms the general idea of a Northwest-Southeast compression in the area.

We can deduce from the attenuation/distance curve that this attenuation is very low and selective in different azimuths.

The obtained horizontal acceleration values with respect to intensity agree with those corresponding to AMBRASEYS (1975) relation. This is also true for the values of the other strong motion data from South Spain corresponding to the 24 June, 1984 earthquake.

REFERENCES

AMBRASEYS, N. N. (1975), *Trends in Engineering Seismology in Europe*, Proc. 5th European Conf. on Earth. Eng. *3*, 39–52, Estambul.

BUFORN, E., UDIAS, A., and MEZCUA, J. (1988), *Seismicity and Focal Mechanisms in South Spain*, Bull. Seismol. Soc. Am. *78*, 2008–2024.

CARREÑO, E., LOPEZ CASDO, C., MARTINEZ SOLARES, J. M., PELAEZ, J. A., TEJEDOR, J. M., and MEZCUA, J. (1988), *Analisis de los primeros acelerogramas registrados en España; terremoto del 24 de junio de 1984* , Informe Especial, Instituto Geográfico Nacional, Madrid.

COULTER, A. W., WALDRON, H. N., and DEVINE, J. F. (1973), *Seismic and Geologic Siting Considerations for Nuclear Facilities*, Proc. V. World Conf. on Earthquake Engineering, Rome.

INSTITUTO GEOGRÁFICO NACIONAL (1991), *Boletín de sismos próximos del año 1989*, 192 pp, Madrid.

MARTIN, A. J. (1984), *Riesgo Sismico en la Peninsula Iberica*, Tesis Doctoral, Escuela de Ingenieros de Caminos, Canales y Puertos, Madrid.

MEZCUA, J. (1982), *Catalogo general de isosistas de la Peninsula Iberica*, Pub. 202, Instituto Geográfico Nacional, Madrid.

MEZCUA, J., and MARTINEZ SOLARES, J. M. (1983), *Seismicidad del área Ibero-Mogrebí*, Instituto Geográfico Nacional, 299, Madrid.

TRIFUNAC, M. D., and BRADY, A. G. (1975), *On the Correlation of Seismic Intensity Scales with Peaks of Recorded Ground Motion*, Bull. Seismol. Soc. Am. *65*, 139–162.

SANZ DE GALDEANO, C. (1988), *Los accidentes y fracturas principales de las Cordilleras Béticas*, Estudios Geol. *39*, 157–165.

(Received February 4, 1991, revised/accepted May 20, 1991)

PAGEOPH, Vol. 136, No. 4 (1991)

0033–4553/91/040405–16$1.50 + 0.20/0

Regional Focal Mechanisms for Earthquakes in the Aegean Area

B. Papazachos,[1] A. Kiratzi[1] and E. Papadimitriou[1]

Abstract—The distribution of the focal mechanisms of the shallow and intermediate depth ($h > 40$ km) earthquakes of the Aegean and the surrounding area is discussed. The data consist of all events of the period 1963–1986 for the shallow, and 1961–1985 for the intermediate depth earthquakes, with $M_s \geq 5.5$. For this purpose, all published fault plane solutions for each event have been collected, reproduced, carefully checked and if possible improved accordingly. The distribution of the focal mechanisms of the earthquakes in the Aegean declares the existence of thrust faulting following the coastline of southern Yugoslavia, Albania and western Greece extending up to the island of Cephalonia. This zone of compression is due to the collision between two continental lithospheres (Apulian-Eurasian). The subduction of the African lithosphere under the Aegean results in the occurrence of thrust faulting along the convex side of the Hellenic arc. These two zones of compression are connected via strike-slip faulting observed at the area of Cephalonia island. The P axis along the convex side of the arc keeps approximately the same strike throughout the arc (210° NNE-SSW) and plunges with a mean angle of 24° to southwest. The broad mainland of Greece as well as western Turkey are dominated by normal faulting with the T axis striking almost NS (with a trend of 174° for Greece and 180° for western Turkey). The intermediate depth seismicity is distributed into two segments of the Benioff zone. In the shallower part of the Benioff zone, which is found directly beneath the inner slope of the sedimentary arc of the Hellenic arc, earthquakes with depths in the range 40–100 km are distributed. The dip angle of the Benioff zone in this area is found equal to 23°. This part of the Benioff zone is coupled with the seismic zone of shallow earthquakes along the arc and it is here that the greatest earthquakes have been observed ($M_s \sim 8.0$). The deeper part (inner) of the Benioff zone, where the earthquakes with depths in the range 100–180 km are distributed, dips with a mean angle of 38° below the volcanic arc of southern Aegean.

Key words: Fault plane solutions, Aegean area.

1. Introduction

The Aegean and the surrounding area is considered to be one of the most seismically active regions of the world. Destructive shallow earthquakes with M_s values up to 7.8, and intermediate depth earthquakes, with even larger magnitudes (~ 8.0), have repeatedly struck many sites of this area. Therefore, a good knowledge of the geodynamics of the region is of primary importance from both the theoretical and practical point of view.

The most prominent features of tectonic origin are, from south to north, the *Mediterranean Ridge*, a compressional submarine accretionary prism of material

---

[1] Geophysical Laboratory, University of Thessaloniki, 540 06 Thessaloniki, Greece.

which extends from the Ionian Sea to Cyprus and follows the trend of the Hellenic arc, the *Hellenic trench* with a maximum water depth of 5 km, the *Hellenic arc*, which consists of the outer sedimentary arc and the inner volcanic arc, and finally the *back-arc Aegean* area, which includes the Aegean Sea, the mainland of Greece, Albania, and south Yugoslavia, south Bulgaria and western Turkey.

In this paper, reliable fault plane solutions are used to derive the pattern of the active stress field in the broader Aegean area. Many focal mechanisms have been published for the earthquakes of the region (McKenzie, 1972, 1978; Ritsema, 1974; Drakopoulos and Delibasis, 1982; Udias *et al.*, 1989 among others). It is often the case to have many controversial solutions for the same event. It was one of the scopes of this paper to gather all published fault plane solutions, check the data and finally adopt one plausible mechanism for any one event.

2. The Data

The earthquakes of the period 1963–1986 for the shallow and of 1961–1985 for the intermediate ($h > 40$ km) depth, were identified and depicted from the catalogue of Comninakis and Papazachos (1986). We start from the early 1960's because of the installation and operation of the World Wide Standard Seismograph Network (WWSSN) at that time. For the earthquakes thus collected, all published fault plane solutions were gathered for each event. We preferred to take into consideration only those that had the polarity data also published on the plots. For all these we reproduced the solution, carefully checked the data and the possibility of any refinement and finally adopted one solution that best fitted the data. The focal parameters of most of the recent strong earthquakes in the Aegean have been determined by waveform modelling (Soufleris and Stewart, 1981; Anderson and Jackson, 1987; Papadimitriou, 1988; Kiratzi and Langston, 1989, 1991; Ioannidou, 1989; Taymaz *et al.*, 1990; Kiratzi *et al.*, 1987, 1991).

Table 1 lists the shallow depth ($h \le 40$ km) earthquakes, that is, the date of occurrence, the origin time, the epicentral coordinates, the surface wave magnitude, the focal depth in km, the azimuth and plunge of the maximum compressional, *P* and the maximum tensional *T*, axes, the strike, dip and rake of the two nodal planes and finally the references of the published fault plane solutions on which they were based. In each case, nodal plane 1 is considered to be the fault plane based on the existence of surface expression, if any, on the distribution of aftershocks, on macroseismic data, and on our general knowledge of the seismotectonics of the region. In cases where no other evidence was available, the nodal plane with the smallest dip angle was chosen as the fault plane.

This is based upon the suggestion that in the thrust zone along the convex side of the Hellenic arc the nodal planes which dip to the inner (Aegean) side of the arc are the fault-planes (Papazachos and Delibasis, 1969; Papazachos and

COMNINAKIS, 1969; MCKENZIE, 1970, 1972) and that these planes have a lower dip angle than the auxiliary planes (MCKENZIE, 1978).

Table 2 lists fault plane solutions of intermediate depth events of the period 1961–1985 which occurred in the southern Aegean area. It is seen that these events have magnitudes smaller than 7. In the past, however, strong ($M \approx 8.0$) earthquakes of intermediate focal depth have occurred there (PAPAZACHOS and COMNINAKIS, 1971; PAPAZACHOS and PAPAZACHOS, 1989; PAPAZACHOS, 1990).

3. The Distribution of Focal Mechanisms

The fault plane solutions of the events listed in Table 1 are illustrated in Figure 1. The black quadrants denote compressional first motion, while the blank ones, denote dilatational first motion. The date of occurrence of the corresponding earthquake is also shown in this figure.

3.1. Yugoslavia-Albania-Western Greece

The distribution of the focal mechanisms of the shallow events (Figure 1), shows the existence of thrust faulting along the Dalmatian coast of Yugoslavia, where the most recent seismicity is expressed by the 1979 Monte Negro sequence. The mainshock and the largest aftershock of this sequence show almost pure thrust faulting with nodal planes striking parallel to the coast (BOORE et al., 1981). This belt of thrust faulting continues south along the coastal region of western Albania, as well as along the westernmost part of the mainland of Greece (events of Oct. 29, 1966 and Oct. 13, 1969). These two events correspond to thrust faulting with left lateral strike-slip motion. The strike of the P axis along the coast of Yugoslavia-Albania-west Greece is about $222° \pm 31°$ and its plunge is about $28° \pm 18°$. The above mentioned zone of thrust faulting is due to the collision of two continental lithospheres (Apulian-Eurasian). There is no evidence of subduction in this part.

3.2. Hellenic Arc

3.2.1. Shallow seismicity. Thrusting continues south, along the coastal regions of northwestern Greece up to the island of Cephalonia. The most recent activity that struck this area of the Ionian islands was expressed by the January 17, 1983 sequence. The mainshock and the largest aftershock revealed dextral strike-slip movement on NE-SW striking nodal planes, which is in agreement with the distribution of the aftershocks (SCORDILIS et al., 1985; SIPKIN 1986; KIRATZI and LANGSTON, 1991). This belt of dextral strike-slip faulting is considered as the westward termination of the Hellenic subduction zone.

Along the convex side of the Hellenic arc, the distribution of focal mechanisms reveals the existence of low angle thrust faulting, as expected, with the shallow

Table 1

The parameters of the focal mechanism of the shallow ($h \leq 40$ km) earthquakes of the Aegean and surrounding area. The data cover the period 1963–1986 with $M_s \geq 5.5$ (Event no. 1 is included due to its large magnitude. The symbol n at the depth means that the depth is in the range of 5–15 km)

No.	Date	Origin Time	Epicentral Coordinates $\phi°N$	$\lambda°E$	Depth h(km)	M_s	P trend°	P plunge°	T trend	T plunge	Nodal plane 1 strike	dip	rake	Nodal plane 2 strike	dip	rake	Ref.
1	1956, July 09	03:11:40	36.7	25.8	n	7.5	155	85	335	5	65	40	-90	245	50	90	1
2	1963, July 26	04:17:12	42.0	21.4	5	6.1	262	25	172	2	322	73	-20	38	70	-163	2
3	1963, Sep. 18	16:58:08	40.8	29.1	n	6.3	177	65	09	26	277	72	94	109	22	76	2
4	1963, Dec. 16	13:47:53	37.0	21.0	15	5.9	195	30	15	60	286	16	90	106	74	90	3
5	1964, Oct. 06	14:31:23	40.3	28.2	n	6.9	148	85	07	03	273	46	-95	101	44	87	4
6	1965, Mar. 09	17:57:54	39.3	23.8	n	6.1	87	0	357	8	40	90	6	310	86	-180	2
7	1965, Apr. 05	03:12:55	37.7	22.0	n	6.1	82	35	178	10	226	57	-159	126	74	-35	2
8	1965, Apr. 27	14:09:06	35.6	23.5	n	5.7	83	71	285	21	22	27	81	191	65	-101	5
9	1965, June 13	20:01:51	37.8	29.3	n	5.6	347	73	169	17	259	38	-90	79	62	-90	4
10	1965, July 06	03:18:42	38.4	22.4	n	6.3	357	61	179	30	270	14	-90	88	76	-90	2
11	1966, Feb. 05	02:01:45	39.1	21.7	8	6.2	163	67	01	19	103	23	75	267	65	98	6
12	1966, May 09	00:42:53	34.4	26.4	16	5.8	205	05	25	85	295	40	90	115	50	90	2
13	1966, Oct. 29	02:39:25	38.9	21.1	20	6.0	277	21	154	58	335	27	132	205	70	63	6
14	1967, Mar. 04	17:58:09	39.2	24.6	8	6.6	320	77	199	05	98	54	-107	302	42	-70	2
15	1967, May 01	07:09:02	39.5	21.2	11	6.4	133	80	280	11	02	36	-100	195	55	-83	4
16	1967, Nov. 30	07:23:50	41.4	20.4	n	6.3	139	76	283	14	0	33	-107	199	58	-79	6
17	1968, Feb. 19	22:45:42	39.4	24.9	9	7.1	83	03	173	07	217	86	175	310	82	4	7
18	1968, Mar. 28	07:39:59	37.8	20.9	6	5.9	231	20	355	59	354	34	137	122	67	63	6
19	1968, May 30	17:40:26	35.4	27.9	n	5.9	202	20	18	70	293	25	90	110	76	90	2
20	1968, Dec. 05	07:52:11	36.6	26.9	n	6.0	354	85	177	05	86	50	-90	266	40	-90	2
21	1969, Jan. 14	23:12:06	36.1	29.2	n	6.2	190	18	05	70	282	25	95	95	75	87	2
22	1969, Mar. 03	00:59:10	40.1	27.5	6	6.0	344	07	79	45	221	64	41	112	54	149	2
23	1969, Mar. 23	21:08:42	39.1	28.5	9	6.1	259	71	358	04	70	46	-128	288	51	63	2
24	1969, Mar. 25	13:21:34	39.2	28.4	n	6.0	257	81	11	08	90	40	-105	290	51	-78	2
25	1969, Mar. 28	01:48:29	38.5	28.5	8	6.6	12	73	191	17	281	28	-90	101	62	-90	2
26	1969, Apr. 03	22:12:22	40.7	20.0	21	5.8	54	16	233	75	143	30	90	323	60	90	21
27	1969, Apr. 06	03:49:34	38.5	26.4	16	5.9	09	75	189	15	280	30	-90	100	60	-90	2
28	1969, Apr. 16	23:21:06	35.2	27.7	35	5.5	197	16	347	71	301	30	109	95	60	80	2
29	1969, June 12	15:13:31	34.4	25.0	19	6.1	192	17	340	72	294	29	105	100	61	80	2
30	1969, July 08	08:09:13	37.5	20.3	10	5.9	243	30	46	61	354	18	115	95	74	81	3
31	1969, Oct. 13	01:02:31	39.8	20.6	8	5.8	194	31	337	41	340	30	160	147	80	62	6
32	1970, Mar. 28	21:02:23	39.2	29.5	10	7.1	09	76	189	14	280	30	-90	90	60	-90	4
33	1970, Mar. 28	23:11:43	39.1	29.6	10	5.2	219	73	359	15	73	32	-109	277	60	-78	8
34	1970, Apr. 08	13:50:28	38.3	22.6	10	6.2	357	75	186	23	278	20	85	90	70	94	8
35	1970, Apr. 16	10:42:22	39.0	29.9	n	5.7	29	75	189	16	273	30	-99	103	59	-85	8
36	1970, Apr. 19	13:29:36	39.0	29.8	18	6.0	191	69	12	21	102	23	90	282	67	90	8
37	1970, Apr. 23	09:01:27	39.1	28.6	28	5.6	325	84	172	05	265	40	-83	78	50	-95	8
38	1970, Aug. 19	02:01:52	41.1	19.8	21	5.4	73	65	253	25	343	20	90	163	70	90	2

No.	Date	Origin time	Lat.	Long.	h	M	P az	P pl	T az	T pl	NP1 strike	NP1 dip	NP1 slip	NP2 strike	NP2 dip	NP2 slip	Ref.
39	1971, May 12	06:25:15	37.6	29.7	10	6.2	160	83	338	05	68	40	-90	247	50	-90	4
40	1971, May 12	10:10:38	37.6	29.7	10	5.6	161	59	343	21	73	14	-90	253	76	-90	4
41	1971, May 12	12:57:25	37.6	29.6	10	5.7	137	65	347	26	79	22	-72	241	70	-97	4
42	1971, May 25	05:43:26	39.0	29.7	10	6.1	243	82	13	07	97	40	-101	288	50	-81	8
43	1972, Mar. 14	14:05:47	39.3	29.5	n	5.6	231	84	05	07	101	40	-101	281	50	-82	8
44	1972, May 04	21:39:57	35.1	23.6	40	6.5	219	27	39	63	308	18	90	129	72	90	9
45	1972, Sep. 17	14:07:15	38.3	20.3	8	6.3	258	37	07	25	46	66	-174	313	84	49	3
46	1973, Jan. 05	05:49:18	35.8	21.9	n	5.6	218	15	46	74	306	30	82	136	60	93	3
47	1973, Nov. 04	15:52:13	38.9	20.5	8	5.8	235	08	349	76	348	40	109	133	51	78	8
48	1973, Nov. 29	10:57:44	35.2	23.8	19	6.0	226	35	44	55	316	10	90	137	80	90	6,8
49	1975, Mar. 27	05:15:08	40.4	26.1	15	6.6	259	56	157	10	41	60	-128	279	46	42	8
50	1976, May 11	16:59:45	37.4	20.4	16	6.5	237	35	57	55	327	12	90	147	78	90	8
51	1976, June 12	00:59:18	37.5	20.6	8	5.8	206	25	26	65	297	20	114	117	70	90	3
52	1977, Aug. 18	09:27:41	35.3	23.5	38	5.6	197	44	29	56	270	12	90	114	79	90	6
53	1977, Sep. 11	23:19:19	34.9	23.0	19	6.3	229	16	349	74	320	30	64	140	60	96	10
54	1978, May 23	23:34:11	40.7	23.2	6	5.8	261	70	173	1	277	49	70	62	46	90	4
55	1978, June 20	20:03:21	40.8	23.2	6	6.5	267	75	46	01	278	46	90	69	48	-116	11
56	1979, Apr. 15	06:19:41	42.0	19.0	4	5.8	230	31	16	57	318	12	90	138	76	-110	11
57	1979, May 15	06:59:23	34.6	24.5	35	5.7	184	29	56	59	253	17	65	100	75	90	12
58	1979, May 24	17:23:18	42.2	18.8	5	6.3	240	20	5	57	330	22	90	150	70	97	12
59	1979, June 14	11:44:45	38.8	26.6	15	5.9	102	61	296	7	121	42	-50	253	52	90	10
60	1979, July 23	11:41:55	35.5	26.4	15	5.5	56	56	345	17	61	35	-40	183	70	-122	12
61	1980, July 09	02:10:20	39.3	22.9	10	5.6	101	83	352	04	82	42	79	247	50	-120	2
62	1980, July 09	02:11:57	39.3	22.9	10	6.5	172	85	351	05	81	40	-90	261	50	-101	22
63	1980, July 09	02:35:52	39.2	22.6	10	6.1	171	85	176	05	81	40	90	261	50	90	13
64	1981, Feb. 24	20:53:37	38.2	23.0	10	6.7	277	69	171	07	290	44	-60	70	54	90	13
65	1981, Feb. 25	02:35:54	38.2	23.1	8	6.4	57	81	57	07	252	43	-104	89	50	-116	13
66	1981, Mar. 04	21:58:07	38.2	23.3	8	6.4	289	85	147	08	60	50	-90	240	40	80	14
67	1981, Dec. 19	14:10:51	39.2	25.2	8	7.2	259	25	352	09	37	67	-166	303	77	86	15
68	1981, Dec. 27	17:39:13	39.2	24.9	8	6.5	77	06	167	03	212	85	-174	123	84	22	14
69	1982, Jan. 18	19:27:25	39.8	24.4	9	7.0	102	13	204	43	235	85	153	343	70	06	16
70	1982, Aug. 17	22:22:20	33.7	22.9	39	6.4	129	02	225	78	230	45	109	20	48	43	16
71	1983, Jan. 17	12:41:30	38.1	20.2	9	7.0	263	25	12	37	40	45	168	140	82	72	16
72	1983, Mar. 23	23:51:05	38.2	20.3	7	6.2	254	12	358	19	29	68	174	123	74	46	10
73	1983, July 05	12:01:27	40.3	27.2	10	6.1	109	32	18	02	248	70	-155	149	66	22	17
74	1983, Aug. 06	15:43:52	40.0	24.7	8	6.8	272	04	02	05	48	83	178	136	88	07	17
75	1984, June 21	10:43:46	35.4	23.3	39	6.2	213	22	10	67	305	24	104	110	68	84	18
76	1985, Apr. 21	08:49:42	35.7	22.2	35	5.6	193	10	60	75	269	36	71	112	56	103	19
77	1986, Sep. 13	17:24:34	37.1	22.2	1	6.0	168	83	285	06	200	50	-81	06	40	-100	23

1 SHIROKOVA (1972), 2 McKENZIE (1972), 3 PAPADIMITRIOU (1990, personal communication), 4 This paper, 5 LYON CAEN et al. (1988) (data from McKENZIE 1972), 6 ANDERSON and JACKSON (1987), 7 KIRATZI et al. (1987, 1991), 8 McKENZIE (1978), 9 KIRATZI and LANGSTON (1989), 10 TAYMAZ et al. (1990), 11 SOUFLERIS and STEWART (1981), 12 BOORE et al. (1981), 13 PAPAZACHOS et al. (1983), 14 PAPAZACHOS et al. (1984b), 15 JACKSON et al. (1982), 16 PAPAZACHOS et al. (1984c), 17 SCORDILIS et al. (1985), 18 DZIEWONSKI et al. (1984), 19 ROCCA et al. (1985), 20 PAPAZACHOS et al. (1988), 21 RITSEMA (1974), 22 EKSTRÖM and ENGLAND (1989), 23 NEIS.

Table 2

The parameters of the focal mechanisms of the intermediate depth ($h > 40$ km) earthquakes occurred at the southern Aegean area. The data cover the period 1961–1985 with $M_s \geq 5.5$

No.	Date	Origin Time	Epicentral Coordinates φ°N	λ°E	Depth h(km)	M_s	P trend[a]	P plunge[a]	T trend[a]	T plunge[a]	Nodal plane 1 strike[a]	dip[a]	rake[a]	Nodal plane 2 strike[a]	dip[a]	rake[a]	Ref.
1	1961, May 23	02:45:20	36.6	28.5	70	6.4	162	13	291	71	270	35	115	60	59	74	1
2	1962, Aug. 28	10:59:56	37.8	22.9	95	6.8	355	03	90	65	241	51	58	107	49	124	2
3	1964, July 17	02:34:27	38.0	23.6	155	6.0	218	24	111	49	267	36	30	153	74	122	2
4	1965, Mar. 31	09:47:31	38.6	22.4	78	6.8	216	29	67	57	286	17	60	136	73	102	3
5	1965, Apr. 09	23:57:02	35.1	24.3	51	6.1	251	17	352	43	23	56	158	127	74	42	3
6	1965, Nov. 28	05:26:05	36.1	27.4	73	6.0	208	30	329	42	350	30	142	89	84	55	3
7	1972, Sep. 13	04:13:20	38.0	22.4	75	6.3	357	21	115	42	235	76	48	131	47	161	4
8	1975, Sep. 22	00:44:56	35.2	26.3	64	5.5	267	19	164	37	310	50	17	209	75	131	5
9	1977, Nov. 28	02:59:10	36.0	27.8	81	5.8	110	17	353	57	166	38	42	40	68	120	4
10	1979, June 15	11:34:17	34.9	24.2	40	5.6	253	28	35	56	21	23	141	150	75	70	5
11	1983, Mar. 19	21:41:42	35.0	25.3	67	5.7	289	02	19	60	43	51	139	162	60	46	5
12	1983, Sep. 27	23:59:39	36.7	26.9	160	5.6	147	28	305	59	261	20	122	49	73	80	4
13	1985, Sep. 27	16:39:49	34.5	26.6	40	5.5	78	02	348	17	125	77	09	31	80	168	5

1 RITSEMA (1974), 2 McKENZIE (1972), 3 This paper, 4 KARACOSTAS (1988), 5 TAYMAZ *et al.* (1990).

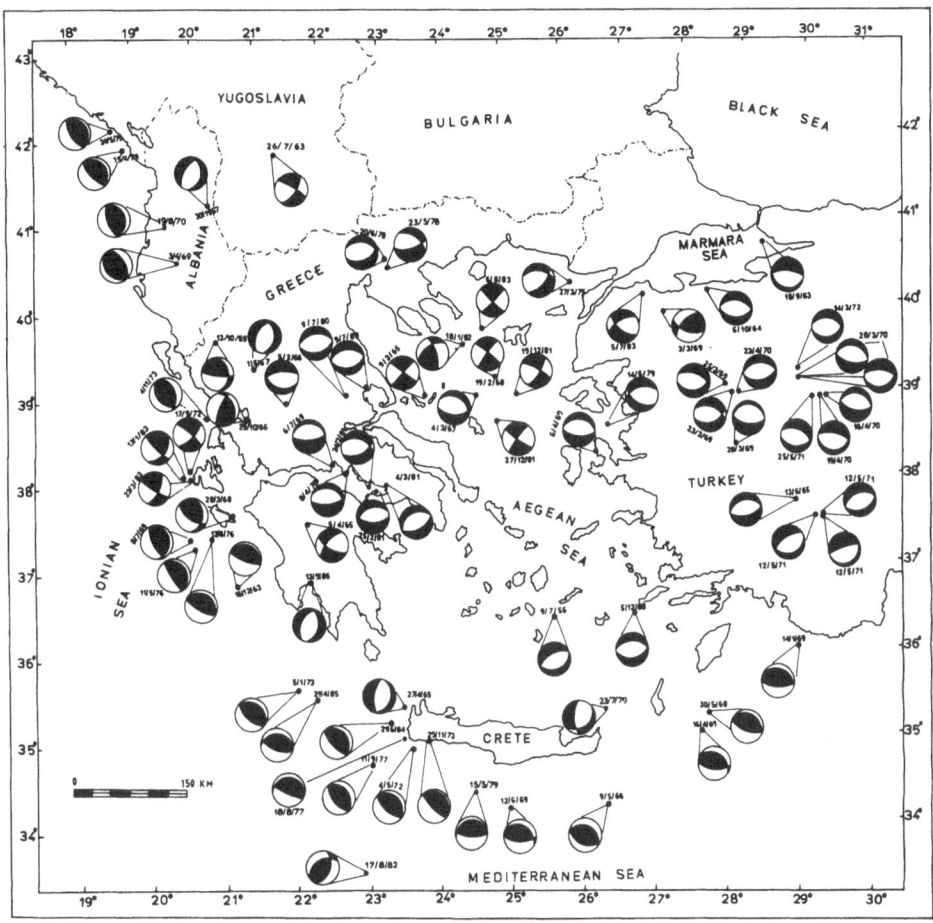

Figure 1

Fault plane solutions of the shallow earthquakes of the Aegean and surrounding area. The data cover the period 1963–1986 with $M_s \geq 5.5$. (The focal mechanism of the earthquake of July 9, 1956 is also shown, due to its large magnitude.) The black quadrants denote compression, while the blank quadrants denote dilatation. The date of occurrence of each event is also shown in this figure.

dipping plane considered as the fault plane (MCKENZIE, 1978; PAPAZACHOS et al., 1984a). Figure 2 shows a lower hemisphere projection of the P axes of the events which occurred along the convex side of the arc. The trend of the P axis is normal to the arc at its western part and keeping the same trend tends to become parallel to the arc at its eastern part. The mean azimuth of the P axis is $210° \pm 17°$ and the mean plunge is $24° \pm 10°$.

3.2.2. Intermediate depth seismicity (h > 40 km). Figure 3 shows the focal mechanisms of the events which occurred in the southern Aegean area, listed in Table 2. All solutions show thrust faulting with considerable strike-slip component. The

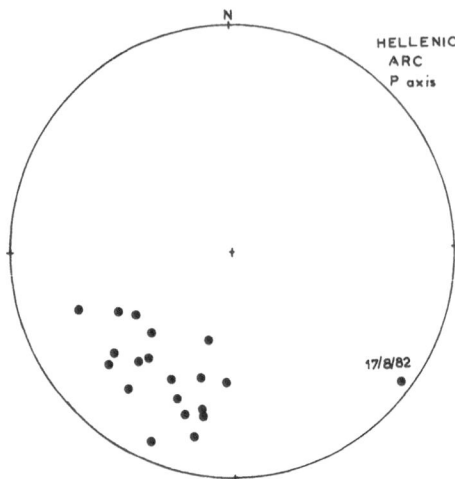

Figure 2
Lower hemisphere projection of the P axes of the shallow events ($h \leq 40$ km) which occurred along the convex side of the Hellenic arc.

earthquake with the largest known focal depth ever observed in this area, is the one which occurred on April 10, 1976 with $M_s = 4.3$, epicentral coordinates 37.31°N, 24.94°E and focal depth 183 km.

Figure 4 shows a cross section along a line perpendicular to the trend of the Hellenic arc at Crete island. In this figure the depth versus the distance from the isodepth of 100 km is plotted. In all cases the depth is determined by waveform modelling (events nos. 4, 5, 8, 10, 11, 13 of Table 2) and is in the range 40–100 km. The slope of the line was found equal to 23°, which corresponds to the slope of the Benioff zone at this depth range. PAPAZACHOS (1990) observed that the moment rate for the shallow seismicity along the external part of the Hellenic arc and for the intermediate depth one is very high and almost the same. Based on this observation he concluded that coupling is taking place on this slightly dipping belt between the eastern Mediterranean lithosphere and the Aegean lithosphere. In this part all the big intermediate depth earthquakes (maximum magnitude ≈ 8.0) occur. The slope of the Benioff zone, using the data from the deeper events, has been previously determined and found equal to 38° (PAPAZACHOS, 1990).

Figure 5 shows a lower hemisphere projection of the slip vectors (A axes) and of the T axes of the earthquakes with shallow depths and with depths in the range 40–100 km. Black circles and black triangles denote the A and T axes, respectively, of the shallow events. Open circles and open triangles denote the A and T axes, respectively, of the events in the depth range 40–100 km, that is of those that form the shallow dipping Benioff zone. This zone is also plotted in Figure 5, with the dashed line. All the azimuths of the A and T axes have been measured from the direction of the Benioff zone at the epicenter of the corresponding event. (The data

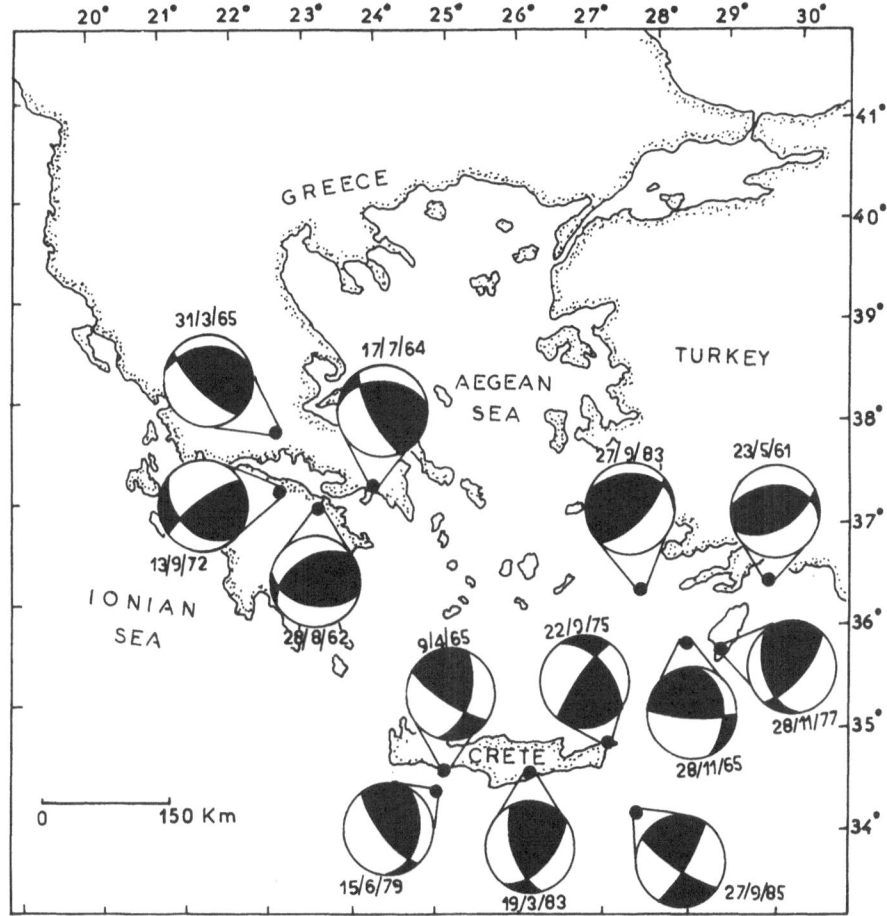

Figure 3
Fault plane solutions of the intermediate depth earthquakes ($h > 40$ km) occurred at the southern
Aegean area. The data cover the period 1961–1985. The black quadrants denote compression while the
blank quadrants denote dilatation.

from the event no. 8 in Table 2 are not plotted in Figure 5.) It is observed from this
figure, that it is the A axes of both the shallow seismicity and the seismicity which
occurred in the depth range 40–100 km, that follow the dip direction of the Benioff
zone (and hence the subducted slab) and not the T axes. This observation has been
previously supported by PAPAZACHOS (1977, 1990) and differs from the results of
ISACKS and MOLNAR (1971), TAYMAZ et al. (1990), who state that the T axes are
aligned down the dip of the subducting lithosphere. The second steepest part of the
Benioff zone (dip angle 38°) is simply the result of the subduction of the front part
of the eastern Mediterranean lithosphere without any coupling with the Aegean
lithosphere. The low seismicity of this inner zone ($M_s \sim 7.0$) compared to the outer
zone is in accordance with this observation.

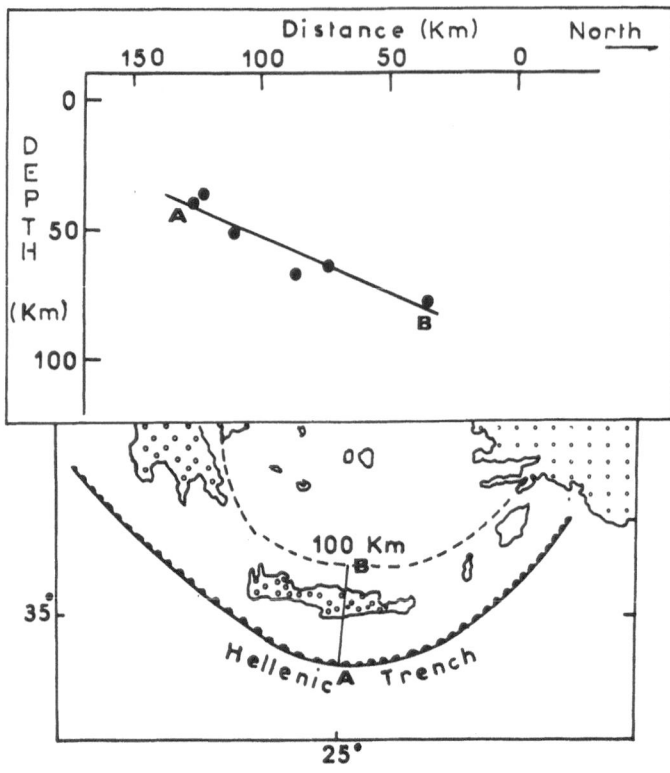

Figure 4

Plot of the depth versus distance from the isodepth of 100 km, along a line perpendicular to the trend of the Hellenic arc at Crete. The earthquakes with depths in the range 41–100 km are plotted. In all cases the depth has been determined by waveform modelling.

3.3. *Mainland of Greece-Western Turkey (Normal Faulting)*

In the mainland of Greece, as well as western Turkey, normal faulting is observed in mainly EW striking nodal planes. The study of the recent seismic sequences (THESSALONIKI, 1978; MAGNESIA, 1980; CORINTH, 1981) has shown that the normal faults are in most cases listric faults occurring at very shallow depths (≤ 10 km).

Figure 6 shows a lower hemisphere projection of the T axes of the events with normal faulting which occurred in the mainland of Greece (black circles). The T axis trends almost NS with a mean direction $174° \pm 12°$ and a mean plunge $10° \pm 8°$. The open circles plotted in Figure 6 denote the T axis of the normal fault events that show EW extension. This zone of EW extension which has been first identified by PAPAZACHOS *et al.* (1984a), extends from Albania, the western edge of the Aegean, Peloponnesus, western Crete and probably eastern Crete. These mechanisms which show normal faulting on NS planes are probably responsible for the peninsulas in

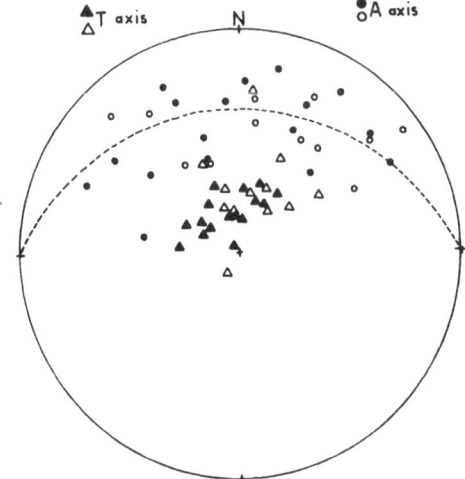

Figure 5

Lower hemisphere projection of the *A* axes (circles) and of the *T* axes (triangles) of the earthquakes which occurred along the convex side of the arc. Black circles and black triangles denote the *A* axes and *T* axes, respectively of the shallow earthquakes ($h < 40$ km), while open circles and open triangles denote the *A* axes and *T* axes, respectively, of the earthquakes in the depth range 41–100 km. The shallow dipping Benioff zone is also plotted on the figure.

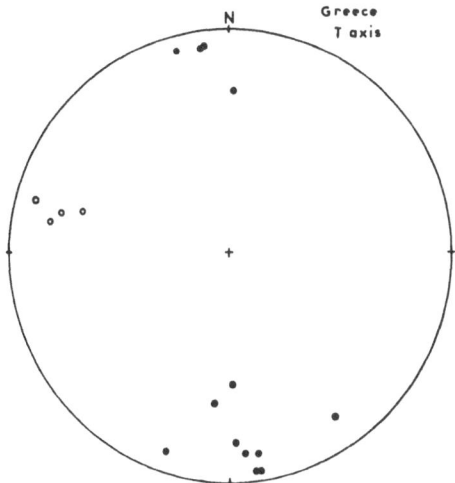

Figure 6

Lower hemisphere projection of the *T* axes of the normal fault earthquakes of the mainland of Greece.

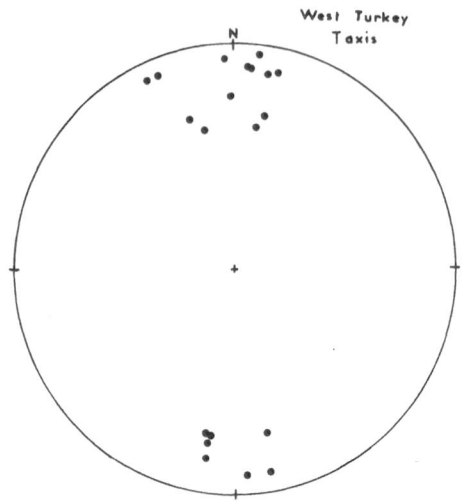

Figure 7
Lower hemisphere projection of the *T* axes of the normal fault earthquakes of western Turkey.

NW and NE corner of Crete (TAYMAZ *et al.*, 1990). The microearthquake data also support the continuation of this zone up to the NE corner of Crete (HATZFELD *et al.*, 1990).

Figure 7 shows a lower hemisphere projection of the *T* axes of the earthquakes which occurred in western Turkey. The extension has an NS trend with a mean azimuth of the *T* axis $180° \pm 12°$ and a mean plunge $12° \pm 7°$.

In the northern Aegean area the focal mechanisms (Figure 1) shows the occurrence of dextral strike-slip faulting (PAPAZACHOS *et al.*, 1984c; ROCCA *et al.*, 1985; KIRATZI *et al.*, 1987, 1991). There are two events, though, which occurred on March 4, 1967 and March 27, 1975 that show normal faulting. The tectonics of this complicated region are probably affected by the continuation of the North Anatolian fault into the Aegean and the NS extension the Aegean itself is undergoing.

4. Conclusions

Figure 8 summarizes the stress pattern in the Aegean and the surrounding area as it is deduced from the seismicity. The occurrence of thrust faulting along the coastal region of Yugoslavia-Albania-western Greece with the *P* axis almost perpendicular to the coastline is shown. Along the island of Cephalonia strike-slip faulting connects this zone of compression with the zone of thrusting along the convex side of the Hellenic arc. The improvement in the depth accuracy of the events along the arc, especially by the work of TAYMAZ *et al.* (1990) has shown that the Benioff zone up to the isodepth of 100 km has a mean dip angle of 23° and then

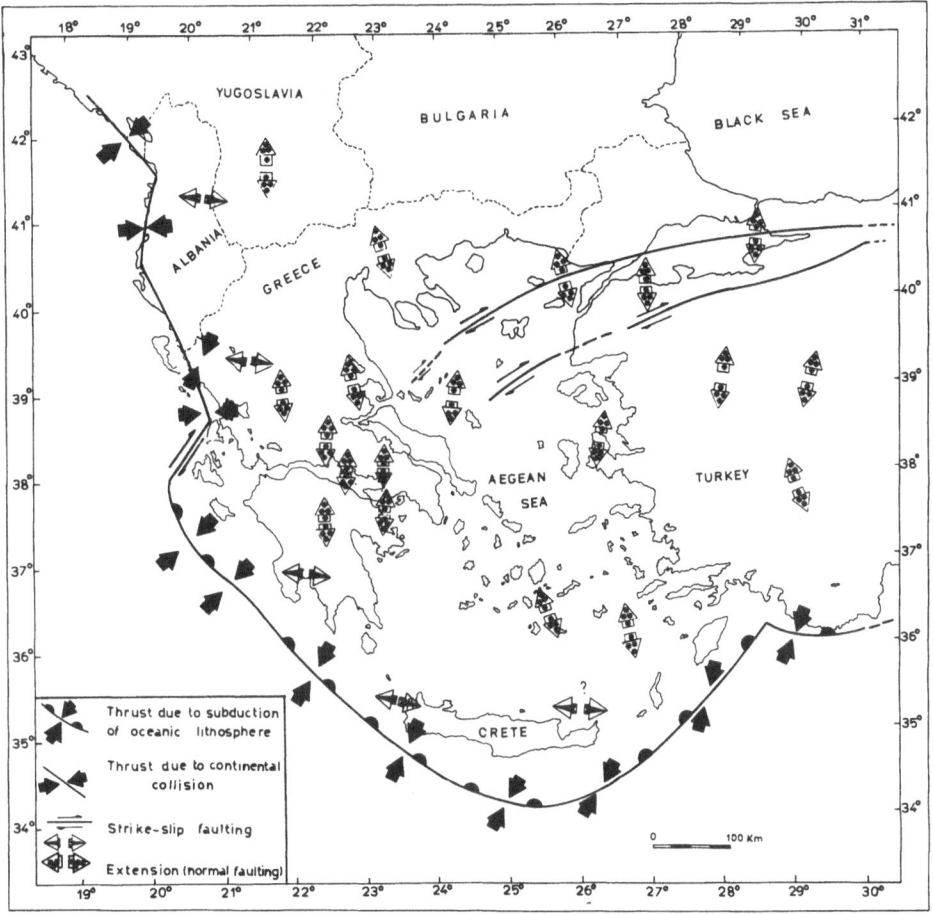

Figure 8
A graph presentation of the stress pattern of the Aegean and the surrounding area as deduced from seismicity.

steepens with a mean dip angle of 38°. The analysis also showed that the A axes are aligned along the subduction and not the T axes as it is the case in other subduction zones. Coupling exists between the eastern Mediterranean lithosphere and the Aegean lithosphere up to the distance of the isodepth of 100 km. The epicenters of the shallow events and of the events in the depth range 40–100 km occur within this coupled area. In this part the strongest earthquakes have been observed and are expected to occur.

The mainland of Greece and western Turkey are characterized by the occurrence of shallow normal faulting in mainly EW striking planes. The direction of the T axes is almost NS. The northern Aegean area seems to be very complicated with the close occurrence of strike-slip events and normal fault events. This is attributed to the control the area is undertaking by the continuation of the North Anatolian fault into the Aegean and the major NS extension the Aegean is undergoing as a whole.

The mechanisms of some earthquakes aligned along a suture zone between the outer zone of thrusting and the inner zone of normal faulting show extension in the EW direction. This zone starts from Albania and with the data existing so far, seems to extend up to the northeastern corner of the island of Crete.

The study of the recent earthquakes of the Aegean with waveform modelling has shown that the seismicity is concentrated almost within the first 10 km of the crust. The intermediate depth seismicity ($h > 40$ km) occurs along the southern part of the Aegean sea.

Acknowledgment

This work was financially supported by the General Secretariat of Research and Technology of Greece under the contract 87AK1/1987.

REFERENCES

ANDERSON, H., and JACKSON, J. (1987), *Active Tectonics of the Adriatic Region*, Geophys. J. R. Astr. Soc. *91*, 937–983.

BOORE, D., SIMS, J., KANAMORI, H., and HARDING, S. (1981), *The Montenegro, Yugoslavia, Earthquake of April 15, 1979: Source Orientation and Strength*, Phys. Earth Planet. Inter. *27*, 133–142.

COMNINAKIS, P., and PAPAZACHOS, B. (1986), *A Catalogue of Earthquakes in Greece and the Surrounding Area for the Period 1901–1985*, Publ. of the Geophys. Lab. *1*, University of Thessaloniki, 167 pp.

DRAKOPOULOS, J., and DELIBASIS, N. (1982), *The Focal Mechanism of Earthquakes in the Major Area of Greece for the Period 1947–1981*, University of Athens, Publ. no. *2*.

DZIEWONSKI, A., FRANZEN, J., and WOODHOUSE, J. (1984), *Centroid-moment Tensor Solutions for July–September, 1983*, Phys. Earth Planet. Int. *34*, 1–8.

EKSTRÖM, G., and ENGLAND, P. (1989), *Seismic Strain Rates in Regions of Distributed Continental Deformation*, J. Geophys. Res. *94*, 10231–10257.

HATZFELD, D., PEDOTTI, G., HATZIDIMITRIOU, P., and MAKROPOULOS, K. (1990), *The Strain Pattern in the Western Hellenic Arc Deduced from a Microearthquake Survey*, Geophys. J. Int. *101*, 181–202.

IOANNIDOU, H. (1989), *Seismic Source Parameters Determined by the Inversion of Body Waves: Greece and the Surrounding Area*, Ph.D. Thesis, University of Athens, 185 pp.

ISACKS, B., and MOLNAR, P. (1971), *Distribution of Stresses in the Descending Lithosphere from a Global Survey of Focal Mechanism Solutions of Mantle Earthquakes*, Rev. Geophys. *9*, 103–174.

JACKSON, J., GAGNEPAIN, J., HOUSEMAN, G., KING, G., PAPADIMITRIOU, P., SOUFLERIS, Ch., and VIRIEUX, J. (1982), *Seismicity, Normal Faulting and Geomorphological Development of the Gulf of Corinth (Greece): The Corinth Earthquakes of February and March 1981*, Earth and Planet. Sci. Lett. *57*, 377–397.

KARACOSTAS, B. (1988), *Fault plane solutions of the earthquakes of the Aegean and the surrounding area*, In Proc. of the 1st Symposium on the Recent Trends in Seismology and Geophysics, Thessaloniki, July 1–3, 1988, pp. 16.

KIRATZI, A., and LANGSTON, C. (1989), *Estimation of Earthquake Source Parameters of the May 4, 1972 Event of the Hellenic Arc by the Inversion of Waveform Data*, Phys. Earth Planet. Inter. *57*, 225–232.

KIRATZI, A., and LANGSTON, C. (1991), *Moment Tensor Inversion of the January 17, 1983 Kefallinia Event of Ionian Islands (Greece)*, Geophys. J. Int. *105*, 529–538.

KIRATZI, A., WAGNER, G., and LANGSTON, C. (1987), *Source Parameters of Some Large Earthquakes in the Aegean Area Determined by Body Waveform Inversion*, AGU Meeting, San Francisco, December 1987 (abstract).

KIRATZI, A., WAGNER, G., and LANGSTON, C. (1991), *Source Parameters of Some Large Earthquakes in Northern Aegean Determined by Body Waveform Inversion*, Pure Appl. Geophys. *105*, 515–527.

LYON-CAEN, H., ARMIJO, R., DRAKOPOULOS, J., BASKOUTAS, J., DELIBASIS, N., GANLON, R., KOUSKOUNA, V., LATOUSSAKIS, J., MAKROPOULOS, K., PAPADIMITRIOU, P., PAPANASTASSIOU, D., and PEDOTTI, G. (1988), *The 1986 Kalamata (South Peloponnesus) Earthquake: Detailed Study of a Normal Fault, Evidence for East-West Extension in the Hellenic Arc*, J. Geophys. Res. *93*, 14967–15000.

McKENZIE, D. P. (1970), *The Plate Tectonics of the Mediterranean Region*, Nature *226*, 239–243.

McKENZIE, D. P. (1972), *Active Tectonics of the Mediterranean Region*, Geophys. J. R. Astr. Soc. *30*, 109–185.

McKENZIE, D. P. (1978), *Active Tectonics of the Alpine-Himalayan Belt: The Aegean Sea and Surrounding Regions*, Geophys. J. R. Astr. Soc. *55*, 217–254.

PAPADIMITRIOU, P. (1988), *Etude de la structure du manteau superieur de l'Europe et modelisation des ondes de volume engendrees par des seismes Egeens*, Ph.D. Thesis, University of Paris VII, 211 pp.

PAPAZACHOS, B. (1977), *A Lithospheric Model to Interpret Focal Properties of Intermediate and Shallow Shocks in Central Greece*, Pure Appl. Geophys. *115*, 655–666.

PAPAZACHOS, B. (1990), *Seismicity of the Aegean and Surrounding Area*, Tectonophysics *178*, 287–308.

PAPAZACHOS, B., and COMNINAKIS, P. E. (1969), *Geophysical Features of the Greek Island Arc and Eastern Mediterranean Ridge*, C. R. Seances de la Conference Réunie à Madrid *16*, 74–75.

PAPAZACHOS, B., and DELIBASIS, N. D. (1969), *Tectonic Stress Field and Seismic Faulting in the Area of Greece*, Tectonophysics *7*, 231–255.

PAPAZACHOS, B., and COMNINAKIS, P. (1971), *Geophysical and Tectonic Features of the Aegean Arc*, J. Geophys. Res. *76*, 8517–8533.

PAPAZACHOS, B., PANAGIOTOPOULOS, D., TSAPANOS, Th., MOUNTRAKIS, D., and DIMOPOULOS, G. (1983), *A Study of the 1980 Summer Seismic Sequence in the Magnesia Region of Central Greece*, Geophys. J. R. Astr. Soc. *75*, 155–168.

PAPAZACHOS, B., KIRATZI, A., HATZIDIMITRIOU, P., and ROCCA, A. (1984a), *Seismic Faults in the Aegean Area*, Tectonophysics *106*, 71–85.

PAPAZACHOS, B., COMNINAKIS, P., PAPADIMITRIOU, E., and SCORDILIS, E. (1984b), *Properties of the February–March 1981 Seismic Sequence in the Alkyonides Gulf of Central Greece*, Annales Geophysicae *2*, 537–544.

PAPAZACHOS, B., KIRATZI, A., VOIDOMATIS, Ph., and PAPAIOANNOU, Ch. (1984c), *A Study of the December 1981-January 1982 Seismic Activity in Northern Aegean Sea*, Boll. Geofis. Teor. Appl. *26*, 101–102.

PAPAZACHOS, B., KIRATZI, A., HATZIDIMITRIOU, P., and KARACOSTAS, B. (1986), *Seismotectonic Properties of the Aegean Area that Restrict Valid Geodynamic Models*, Wegener/Medlas Conference, Athens, May 14–16, 1986, pp. 1–20.

PAPAZACHOS, B., KIRATZI, A., KARACOSTAS, B., PANAGIOTOPOULOS, D., SCORDILIS, E., and MOUNTRAKIS, D. (1988), *Surface Fault Traces, Fault Plane Solution and Spatial Distribution of the Aftershock of the September 13, 1986 Earthquake of Kalamata (Southern Greece)*, Pure Appl. Geophys. *126*, 55–68.

PAPAZACHOS, B., and PAPAZACHOS, C., *The Earthquakes of Greece* (Ziti Publication Co., 1989) pp. 356.

RITSEMA, A. (1974), *The Earthquake Mechanisms of the Balkan Region*, R. Netherl. Meteorol. Inst., De Bilt, Sci. Rep. 74-4.

ROCCA, A., KARAKAISIS, G., KARACOSTAS, B., KIRATZI, A., SCORDILIS, E., and PAPAZACHOS, B. (1985), *Further Evidence on the Strike-slip Faulting of the Northern Aegean Trough Based on Properties of the August–November 1983 Seismic Sequence*, Boll. di Geof. Teor. Appl. *27*, 101–109.

SCORDILIS, E., KARAKAISIS, G., KARACOSTAS, B., PANAGIOTOPOULOS, D., COMNINAKIS, P., and PAPAZACHOS, B. (1985), *Evidence for Transform Faulting in the Ionian Sea: The Cephalonia Island Earthquake Sequence of 1983*, Pure Appl. Geophys. *123*, 388–397.

SHIROKOVA, E., *Stress pattern and probable motion in the earthquake foci of the Asia-Mediterranean seismic belt*, In *Elastic Strain Field of the Earth and Mechanisms of Earthquake Sources*, L. M. Balakina *et al.* (eds), (Nauka, Moscow, 8 1972).

SIPKIN, S. (1986), *Estimation of Earthquake Source Parameters by the Inversion of Waveform Data*: *Global Seismicity*, 1981–1983, Bull. Seismol. Soc. Am. *76*, 1515–1541.

SOUFLERIS, Ch., and STEWART, G. (1981), *A Source Study of the Thessaloniki (N. Greece) 1978 Earthquake Sequence*, Geophys. J. R. Astr. Soc. *67*, 343–358.

TAYMAZ, T., JACKSON, J., and WESTAWAY, R. (1990), *Earthquake Mechanisms in the Hellenic Trench Near Crete*, Geophys. J. Int. *102*, 695–731.

UDIAS, A., BUFORN, E., and RUIZ DE GAUNA, J. (1989), *Catalogue of Focal Mechanisms of European Earthquakes*, Department of Geophysics, Universidad Complutense, Madrid, 274 pp.

(Received February 11, 1991, revised/accepted May 21, 1991)

PAGEOPH, Vol. 136, No. 4 (1991)

0033–4553/91/040421–12$1.50 + 0.20/0

Rates of Crustal Deformation in the North Aegean Trough-North Anatolian Fault Deduced from Seismicity

ANASTASIA A. KIRATZI[1]

Abstract—The rates and configuration of seismic deformation in the North Aegean trough-North Anatolian fault are determined from the moment tensor mechanisms of the earthquakes that occurred within this region. The analysis is based on KOSTROV's (1974) formulation. The fault plane solutions of the earthquakes of the period 1913–1983 with $M_s \geq 6.0$ are used. The focal mechanism of some of the past events (before 1960) is assumed, based on the present knowledge of the seismotectonics as well as on the macroseismic records of the area studied. The analysis showed that the deformation of the northern Aegean is dominated by EW contraction (at a rate of about 15 mm/yr) which is relieved by NS extension (at a rate of about 9 mm/yr). It was also shown that the northern part of North Anatolia (north of 39.7°N parallel) undergoes contraction in the EW direction (at a rate of about 9 mm/yr) and NS extension as the dominant mode of deformation (at a rate of about 5 mm/yr). It may be stated therefore, that the pattern of deformation of the northern Aegean and the northern part of North Anatolian fault is controlled by the NS extension the Aegean is undergoing as a whole, and the dextral strike-slip motion of the North Anatolian fault. The southern part of North Anatolia is undergoing crustal thinning at a rate of 2.3 mm/yr, NS extension (at a rate of 5 mm/yr) as well as EW extension (at a rate of 4 mm/yr), which are consistent with the occurrence of major normal faulting and justify the separation of North Anatolia into two separate subareas.

Key words: Crustal deformation, North Aegean trough, North Anatolia.

1. Introduction

This paper deals with the deformation rates, as they are deduced from seismicity, of the northern Aegean area, in conjuction with the deformation rates of the northern Anatolian fault. Figure 1 illustrates the area in which its deformation pattern is examined here.

The northern Aegean area is dominated by numerous subsiding basins among which the most important are the North Aegean trough and the Marmara Sea. The structure of these basins is produced by major faulting with up to 6–7 km of normal displacement, which is still continuing as the North Aegean trough, the Marmara Sea and the North Anatolian fault are very active seismically (MCKENZIE, 1972, 1978; PAPAZACHOS, 1990). The North Aegean trough is a graben,

[1] Geophysical Laboratory, University of Thessaloniki, 540 06 Thessaloniki, Greece.

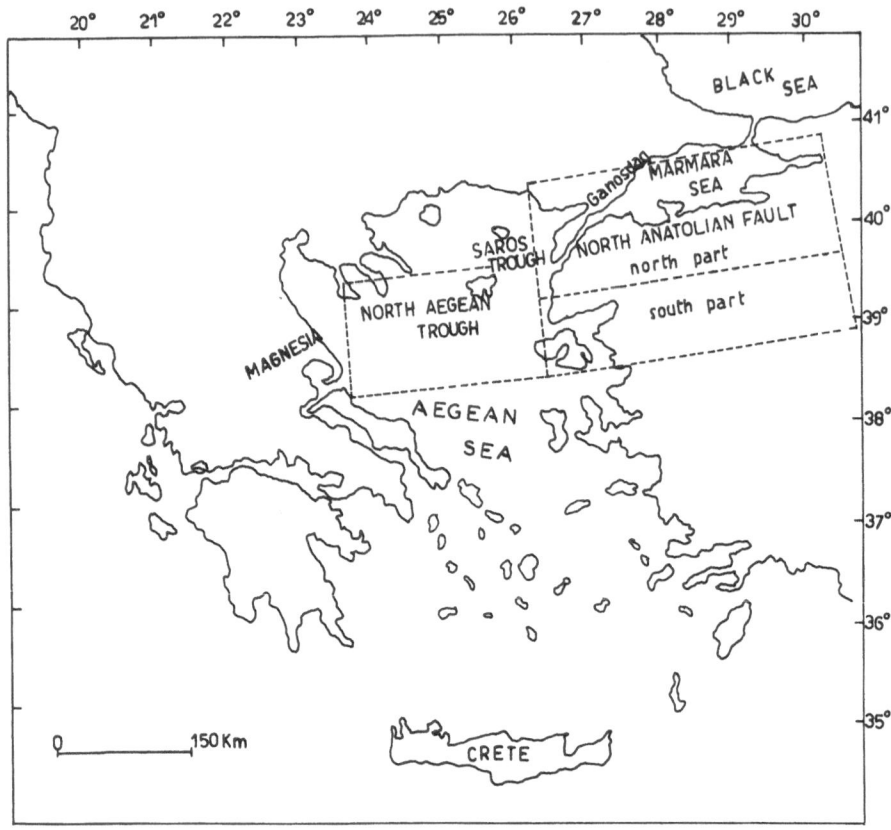

Figure 1
The region of North Aegean and North Anatolian fault the deformation of which is examined. The
dashed lines show the rectangle approximation to the area studied.

1000–1500 m deep, which extends from the Saros trough to the east as far as
Magnesia to the west. This trough and its surrounding basins constitute the
southern margin of the European plate.

The Ganosdag area of active faults joins the North Aegean trough to the
Marmara Sea and establishes a tectonic connection with the North Anatolian fault.
This fault, which marks the deforming zone between Turkey and Eurasia, is
1000 km long and very active. It is a strike-slip right lateral fault which has broken
along most of its length since 1939 (AMBRASEYS, 1970).

The northern Aegean area is affected by a prevailing tectonic regime of
extension coupled with strike-slip motion due to the existence of the North
Anatolian fault, since the uppermost Miocene, which is still active (LYBERIS, 1981).
Large earthquakes have recently occurred in the northern Aegean area (PAPAZA-
CHOS et al., 1984; ROCCA et al., 1985) and the historical record also shows intense

activity in the region (PAPAZAHCOS and PAPAZACHOS, 1989). It is the deformation of this tectonically complex area which is the major concern of this paper. Previous work by JACKSON and McKENZIE (1988) and EYIDOGAN (1988) deals with deformation rates but they use a different set of data applied to a different part of the area selected here.

2. Method of Analysis

The analysis is based on KOSTROV's (1974) formulation according to which, the average strain rate of seismic deformation in a region can be defined as:

$$\dot{\varepsilon}_{ij} = 1/(2\,\mu VT)\left\{\Sigma M_{ij}\right\} \tag{1}$$

where ΣM_{ij} is the sum of symmetric moment tensors of N earthquakes in the seismogenic volume V of the deforming zone of rigidity μ in time T. This formula is applicable to the case in which the margins of the deforming zone are in the far-field and thus the method estimates the rate of irrotational strain due to slip on faults with a variety of orientations.

The area studied is considered as a rectangular deforming region of thickness Z in the vertical (z) direction bounded by sides of lengths X and Y in the two horizontal directions x and y. A coordinate frame is used with the $x =$ north, $y =$ east and $z =$ down directions positive. The x axis is taken to be normal to the average trend of the deforming zone of northern Aegean-northern Anatolian.

The six elements of the seismic moment tensor for each fault plane solution, were obtained, using the relation

$$\mathbf{M}_{ij} = Mo(\mathbf{u}_i \mathbf{n}_j + \mathbf{u}_j \mathbf{n}_i) \tag{2}$$

where $\mathbf{M}_{ij}$ is the moment tensor, Mo is the scalar moment, and $\mathbf{u}$ and $\mathbf{n}$ are unit vectors in the direction of the slip vector and the normal to the fault plane, respectively (AKI and RICHARDS, 1980). Then, following the formulations developed in JACKSON and McKENZIE (1988), and using the above-mentioned coordinate system the components of the rate of deformation were calculated directly from the moment tensors. The extent to which deformation is taken up by movement of material in the x direction is given by the following equation:

$$U_x^x = (1/2\mu ltT) \sum_{n=1}^{N} \mathbf{M}_{11} \tag{3}$$

where $\mu = 3*10^{11}$ dyn/cm^2, l is the length of the deforming zone, T is the time of observations, and t is the thickness of the seismogenic layer.

The extent to which deformation is taken up by movement of material in the y direction is calculated by M_{22}, since

$$U_y^y = (1/2 \, \mu T\alpha t) \sum_{n=1}^{N} M_{22} \qquad (4)$$

where α is now the width of the deforming zone.

So, the rate of thickening in the seismogenic layer is given by the component of the moment tensor M_{33},

$$U_z^z = (1/2 \, \mu T\alpha l) \sum_{n=1}^{N} M_{33}. \qquad (5)$$

As is stated in JACKSON and MCKENZIE (1988) the off-diagonal terms of the moment tensor do not have simple physical meaning. But if the length, l, of the deforming zone is much greater than its width, α, then some components of the tensor do have physical sense. In this respect, the horizontal shear velocity is determined by the equation

$$U_y^x = (1/\mu \tau t l) \sum_{n=1}^{N} M_{12}. \qquad (6)$$

Similarly, the gradients of the horizontal velocity in the vertical direction are given by the equations

$$U_z^x = (1/\mu \alpha l T) \sum_{n=1}^{N} M_{13} \qquad (7)$$

$$U_z^y = (1/\mu \alpha l T) \sum_{n=1}^{N} M_{23}. \qquad (8)$$

The reader is prompted to read the papers by JACKSON and MCKENZIE (1988) and EKSTROM and ENGLAND (1989) for further details on the equations and the limitations in their application.

3. The Data

The data used in this study consist of all the earthquakes with $M_s \geqslant 6.0$ that occurred in the North Aegean-North Anatolian during the period 1913–1983. The catalogue of COMNINAKIS and PAPAZACHOS (1986) was used to identify the events.

The scalar moment was, in most cases, determined by a moment-magnitude relation, determined by KIRATZI et al. (1985) for earthquakes in the Aegean area which is shown in Figure 2. This relation is in very good agreement with the one

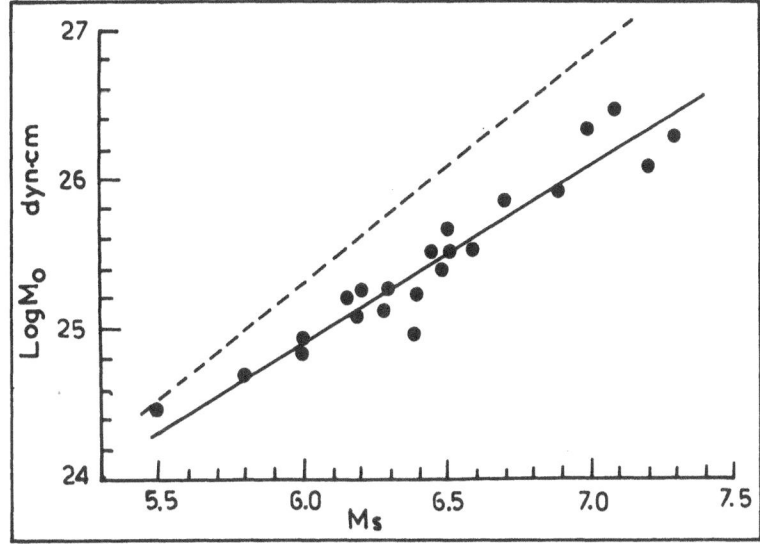

Figure 2

Plot of M_s and Mo values for earthquakes in Greece (KIRATZI *et al.*, 1985). The dashed line shows the global relation determined by DZIEWONSKI and WOODHOUSE (1983).

later proposed by PAPAZACHOS (1989). The relationship of DZIEWONSKI and WOODHOUSE (1983) based on worldwide data, expressed by the dashed line, is also shown for comparison, in this figure.

The northern Aegean area is considered to be rectangular bounded by the geographical coordinates, 38.9°N–23.4°E, 40.1°N–23.4°E, 40.1°N–26.0°E, 38.9°N–26.0°E (see Fig. 1). The North Anatolia is also approximated by a rectangle, bounded by the geographical coordinates 38.9°N–26.0°E, 38.9°N–30.0°E, 40.9°N–30.0°E, 40.9°N–26.0°E. This area of North Anatolia was separated into two seismotectonic units, that is, its northern part, where the Sea of Marmara is located, and into its southern part, south of 39.7°N, where the southern provinces of Anatolia are situated. This was decided after some initial calculations and examination of deformation patterns.

For the data set, thus collected, a great effort was made to obtain the most reliable and well constrained fault plane solutions possible. For the events which occurred after the installation of the Worldwide Standard Seismograph Network (WWSSN) in the early 1960's, there was no major problem. For the past events however, for which there was no focal mechanism published, a mechanism was assumed, based on the present knowledge of the seismotectonics of the region, the macroseismic data if any, and finally on the information given in the book of PAPAZACHOS and PAPAZACHOS (1989). Figure 3 shows the available focal

Table 1

Earthquakes of the period 1913–1983, occurred in the northern Aegean trough with $M_s \geq 6.0$ (see text for more information)

| Date | Epicenter | | M_s | Mo 10^{25} dyn.cm | Fault plane solution | | | Moment tensor elements | | | | | | REF. |
	$\phi°N$	$\lambda°E$			str	dip	rake	M_{11}	M_{22}	M_{12}	M_{13}	M_{23}	M_{33}	
5 Dec., 1923	39.8	23.5	6.4	2.54	(135	45	−90)	1.27	1.27	1.27	0.00	0.00	−2.54	1
9 Mar., 1965	39.3	23.8	6.1	1.10	40	90	180	1.08	−1.08	−0.19	0.00	0.00	0.00	4
4 Mar., 1967	39.2	24.6	6.6	5.23	98	41	−110	4.45	0.42	1.80	0.49	1.43	−4.87	5
19 Feb., 1968	39.4	24.9	7.1	22.42	216	81	173	21.10	−21.00	−6.24	−4.72	0.13	−0.10	6
19 Dec., 1981	39.2	25.2	7.2	22.38	47	77	−167	20.50	−21.10	0.62	0.06	8.06	0.60	6
27 Dec., 1981	38.9	24.9	6.5	3.10	224	88	176	3.08	−3.10	−0.10	−0.23	0.08	0.02	5
18 Jan., 1982	39.8	24.4	7.0	13.49	235	50	153	4.60	−10.63	5.99	−5.30	−5.72	6.03	7
6 Aug., 1983	40.0	24.7	6.8	12.11	50	76	177	11.20	−11.90	1.97	2.18	1.91	0.70	6
				82.37				67.28	−67.12	5.12	−7.52	5.89	−0.16	

Table 2

Earthquakes of the period 1913–1983, occurred in the northern part of North Anatolian with $M_s \geq 6.0$ (see text for more information)

| Date | Epicenter | | M_s | Mo 10^{25} dyn.cm | Fault plane solution | | | Moment tensor elements | | | | | | REF. |
	$\phi°N$	$\lambda°E$			str	dip	rake	M_{11}	M_{22}	M_{12}	M_{13}	M_{23}	M_{33}	
4 Jan., 1935	40.4	27.5	6.4	2.54	(302	36	−90)	1.74	0.68	1.08	−0.67	−0.42	−2.42	4
4 Jan., 1935	40.3	27.5	6.3	1.92	(302	36	−90)	1.31	0.51	0.82	−0.50	−0.31	−1.83	8
18 Mar., 1953	40.0	27.4	7.4	41.11	60	90	180	35.60	−35.60	20.56	0.00	0.00	0.00	4
18 Sep., 1963	40.8	29.1	6.3	1.92	268	70	−125	1.08	−0.07	1.00	1.19	−0.42	−1.01	4
6 Oct., 1964	40.3	28.2	6.9	10.21	302	36	−90	6.98	2.73	4.36	−2.68	−1.67	−9.71	4
27 Mar., 1975	40.4	26.1	6.6	4.43	41	60	−125	3.53	−0.39	−1.86	−0.23	2.20	−3.14	4
5 July, 1983	40.3	27.2	6.1	1.10	149	66	−22	0.90	−0.60	0.57	0.21	−0.54	−0.31	2
				63.23				51.14	−32.54	26.53	−2.16	−1.33	−18.42	

Table 3

Earthquakes of the period 1913–1983, occurred in the southern part of North Anatolian with $M_s \geq 6.0$ (see text for more information)

Date	Epicenter		M_s	M_o 10^{25} dyn.cm	Fault plane solution			Moment tensor elements						REF.
	$\phi°N$	$\lambda°E$			str	dip	rake	M_{11}	M_{22}	M_{12}	M_{13}	M_{23}	M_{33}	
18 Nov., 1919	39.3	26.7	7.0	13.49	(270	45	−90)	13.49	0.00	0.00	0.00	0.00	−13.49	4
2 May, 1928	39.6	29.1	6.2	1.45	(308	35	−90)	0.85	0.52	0.66	−0.39	−0.31	−1.36	8
22 Sep., 1939	39.0	26.9	6.6	4.43	(270	45	−90)	4.43	0.00	0.00	0.00	0.00	−4.43	8
28 Oct., 1942	39.1	27.8	6.0	0.83	(90	40	−104)	0.79	0.00	0.13	0.14	0.15	−0.79	8
15 Nov., 1942	39.4	28.1	6.2	1.45	(90	40	−104)	1.39	0.00	0.23	0.24	0.27	−1.39	8
25 June, 1944	39.0	29.3	6.1	1.10	(308	35	−90)	0.64	0.39	0.50	−0.30	−0.23	−1.03	8
6 Oct., 1944	39.4	26.7	6.9	10.21	(80	45	−90)	9.90	0.31	−1.75	0.00	0.00	−10.21	4
23 Mar., 1969	39.1	28.5	6.1	1.10	112	34	270	0.88	0.14	0.35	0.38	0.15	−1.02	3
25 Mar., 1969	39.2	28.4	6.0	0.83	90	40	−104	0.79	0.00	0.13	0.14	0.15	−0.79	4
28 Mar., 1970	39.2	29.5	7.1	26.25	308	35	−90	15.33	9.35	11.97	−7.09	−5.54	−24.68	4
19 Apr., 1970	39.0	29.8	6.0	0.83	284	66	270	0.58	0.04	0.14	0.54	0.13	−0.62	3
25 May, 1971	39.0	29.7	6.1	1.10	298	55	284	0.96	0.04	0.29	0.25	0.31	−1.00	3
				63.07				50.03	10.79	12.66	−6.09	−4.92	−60.81	

Key to the references of the three tables:
1 AMBRASEYS, N. (1970)
2 DZIEWONSKI, A. et al. (1984)
3 EYIDOGAN, H. (1988)
4 JACKSON, J., and MCKENZIE, D. (1988)
5 IOANNIDOU, E. (1989)
6 KIRATZI, A. et al. (1991)
7 ROCCA, A. et al. (1985)
8 Arbitrarily chosen by the author.

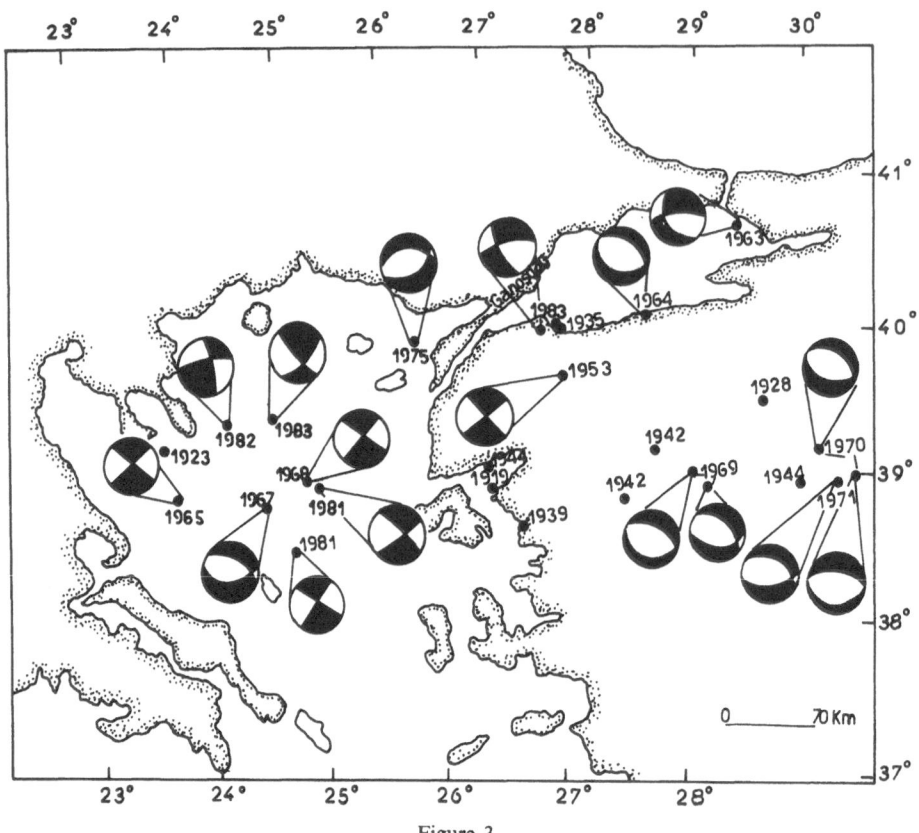

Figure 3

Fault plane solutions of the earthquakes used in this analysis. For those events that a mechanism was assumed, the epicenter and the year of occurrence is only shown on the map.

mechanisms. For those events that a focal mechanism was assumed their epicenter and year of occurrence is only shown in this figure.

Tables 1, 2 and 3 list the earthquakes for every region. These tables contain the data of occurrence, the epicentral coordinates, the surface wave magnitude, the seismic moment, the fault plane solution, the components of the determined moment tensor following AKI and RICHARDS (1980) formulation, and finally the reference from which the fault plane solution was taken. The fault plane solutions that were assumed are listed in parentheses in Tables 1–3. The summed seismic moment tensor elements for each area are also shown in these tables. If the sum of the moment tensor components given in these tables, does not equal zero, as it should, it is due to rounding effects.

The thickness of the seismogenic layer was assumed to be 10 km in all cases, based on waveform modelling and previous work (EYIDOGAN, 1988; EKSTRÖM and ENGLAND, 1989; KIRATZI, 1990; KIRATZI et al., 1991).

4. Rates of Deformation

a. Northern Aegean Trough

The total dimension of the deformed volume of northern Aegean in the NS and EW directions is about 220 km and 125 km, respectively. Assuming a seismogenic layer of 10 km the following strain rate tensor is calculated for the time span of observations, listed in Table 1, 1913–1983 ($T = 60$ yrs):

$$\dot{\varepsilon}_{ij} = \begin{bmatrix} 68.0 & 5.2 & -7.6 \\ 5.2 & -67.8 & 5.9 \\ -7.6 & 5.9 & -0.2 \end{bmatrix} * 10^{-9}/\text{yr}. \tag{9}$$

The dominant components $\dot{\varepsilon}_{11}$ and $\dot{\varepsilon}_{22}$ correspond to extension in the NS direction and compression in the EW direction. The corresponding velocities are equal to 8.5 mm/yr for the extension and 14.9 mm/yr for the EW contraction. The results also show right lateral shear in the EW direction, or left lateral shear in the NS direction, corresponding to about 1.3 mm/yr and thinning of the seismogenic layer, at a rate of only 0.15 mm/yr. For the calculation of the shear motion one has to remember the restrictions regarding the length of the deforming region in comparison with its width as they are previously mentioned. But since in all cases here the length is much greater than the width, of the deforming region we attempt some rough shear motion estimates.

The seismic moment release from the 8 events used, averaged over 60 years, is $1.37 * 10^{25}$ dyn. cm/yr. It was calculated that the contribution of events with magnitudes less than 6.0 have a less than 5% effect on the total seismic moment. The contribution of the largest event of the area (19/2/1968) to the total sum of seismic moment is about 27%.

b. Northern part of North Anatolian Fault

The seismic moment release rate from the 7 events listed in Table 2 and averaged over 49 years is $1.29 * 10^{25}$ dyn. cm/yr. Assuming that the length and the width of the deforming zone are 320 km and 120 km, respectively, then the strain rate tensor is:

$$\dot{\varepsilon}_{ij} = \begin{bmatrix} 45.3 & 23.5 & -1.9 \\ 23.5 & -29.0 & -1.2 \\ -1.9 & -1.2 & -16.3 \end{bmatrix} * 10^{-9}/\text{yr}. \tag{10}$$

It is observed that the dominant components of the tensor are again $\dot{\varepsilon}_{11}$ and $\dot{\varepsilon}_{22}$ corresponding to 5.4 mm/yr of extension in the NS direction and 9.3 mm/yr of contraction in the EW direction. The tensor also reveals that the average right lateral shear motion in the EW direction is 5.6 mm/yr, which is considerably higher than the shear motion of the northern Aegean trough. The thinning of the seismogenic layer corresponds to approximately 0.16 mm/yr.

c. Southern Part of North Anatolian Fault

The seismic moment release rate for the 12 events listed in Table 3, averaged over 52 yrs is $1.21 * 10^{25}$ dyn. cm/yr. The length of the deforming area is 350 km and its width 85 km. The strain rate tensor was calculated to be as follows:

$$\dot{\varepsilon}_{ij} = \begin{bmatrix} 53.8 & 13.6 & -6.6 \\ 13.6 & 11.6 & -5.3 \\ -6.6 & -5.3 & -65.5 \end{bmatrix} * 10^{-9}/\text{yr}. \tag{11}$$

It is seen that the dominant components are $\dot{\varepsilon}_{33}$ and $\dot{\varepsilon}_{11}$ which correspond to about 0.7 mm/yr thinning of the seismogenic layer, consistent with the occurrence of normal faulting and to about 5 mm/yr NS extension across the region. Also observed is a component of EW extension with a rate of about 4 mm/yr. The shear motion in this case corresponds to 2.3 mm/yr of NS left lateral ($\dot{\varepsilon}_{12}$) or EW right lateral ($\dot{\varepsilon}_{21}$) shear across the area.

5. Discussion and Conclusions

In this paper the deformation of the northern Aegean and part of the North Anatolian fault was examined. The analysis was based on KOSTROV's (1974) formulation. The data set consists of all the earthquakes with $M_s \geq 6.0$ covering the period 1913–1983.

If we assume that the greatest uncertainties in the $\mathbf{M}_{ij}$ come from the estimates of Mo (MOLNAR and DENG, 1984) and not from the uncertainties in the strike, dip and rake of each fault, then we must check the errors inherited in its calculation. The equation of KIRATZI et al. (1985) used in the cases where there was no independent estimation of Mo, has an rms residual of $\pm 0.2 M_s$ which is very close to the accuracy of the determination of M_s values (KIRATZI and PAPAZACHOS, 1984; PAPAZACHOS and PAPAZACHOS, 1989). The resulting uncertainty in the estimation of Mo is about a factor of 2, which in turn affects by roughly the same factor, the calculated deformation rates.

Therefore, within these errors limits, the calculated strain tensors indicate that the dominant mode of deformation in the northern Aegean area is associated with NS extension at a rate of about 9 mm/yr and EW contraction at a rate of 15 mm/yr. The northern part of the North Anatolian fault (Marmara area) exhibits the same pattern of deformation. That is NS extension at a rate of about 5 mm/yr and EW compression at a rate of about 9 mm/yr. The right lateral shear motion is estimated to have a rate of 6 mm/yr. The southern part of this same fault exhibits crustal thinning as the dominant mode of deformation, in agreement with the occurrence of major normal faulting, at a rate of 2.3 mm/yr. The NS extension here is estimated to have a rate of 5 mm/yr, the same as farther north. However, EW extension is observed at a rate of 4 mm/yr, which suggests the separation from the Marmara province where EW compression is prevailing. These results are in quite good agreement with those obtained by EYIDOGAN (1988). JACKSON and McKEN-ZIE (1988) estimate the NS extension in the whole Aegean to have a rate of 63 mm/yr, which is by a factor of about 4 greater than the rate obtained here. This difference is mainly attributed to the formula they used to calculate Mo, that is the global formula of DZIEWONSKI and WOODHOUSE (1983), shown by the dashed line in Figure 2. This formula results in Mo values greater by a factor of about 4 than the ones used in this paper. The formula between Mo and M_s proposed by PAPAZACHOS (1989) also calculates Mo values about a factor of 4 smaller than the DZIEWONSKI and WOODHOUSE (1983) corresponding global relation. McKENZIE (1978) estimated, from plate motions, the extension in the Aegean to be 70 mm/yr in the direction SSW. This is probably an upper bound. Nevertheless, the area studied here is very limited and the time span probably not representative of longer periods of activity, in order to conclude whether aseismic deformation is invoked in the upper crust, which seems to be the case from the results of this paper, given the method of analysis followed and the data used.

Acknowledgments

The author is particularly indebted to Prof. Papazachos of the University of Thessaloniki for his critical reading of the manuscript and his very helpful suggestions.

This work was financially supported by the General Secretariat of Research and Technology of Greece under the contract 87AK1/1987.

REFERENCES

AKI, K., and RICHARDS, P., *Quantitative Seismology; Theory and Methods* (2 vols.), (W. H. Freeman, San Francisco 1980).

AMBRASEYS, N. (1970), *Some Characteristic Features of the North Anatolian Fault Zone*, Tectonophysics 9, 143–165.

COMNINAKIS, P., and PAPAZACHOS, B. (1986), *A Catalogue of Earthquakes in Greece and Surrounding Area for the Period 1901–1985*, Publ. Geophys. Lab., Univ. of Thessaloniki, 1, 167 pp.

DZIEWONSKI, A., and WOODHOUSE, J. (1983), *An Experiment in Systematic Study of Global Seismicity: Centroid-moment Tensor Solutions for 201 Moderate and Large Earthquakes in 1981*, J. Geophys. Res. 88, 3247–3271.

DZIEWONSKI, A., FRIEDMAN, A., and WOODHOUSE, J. (1984), *Centroid-moment Tensor Solutions for July–September, 1983*, Phys. Earth Planet. Inter. *34*, 1–8.

EKSTRÖM, G., and ENGLAND, P. (1989), *Seismic Strain Rates in Regions of Distributed Continental Deformation*, J. Geophys. Res. *94*, 10231–10257.

EYIDOGAN, H. (1988), *Rates of Crustal Deformation in Western Turkey as Deduced from Major Earthquakes*, Tectonophysics *148*, 83–92.

JACKSON, J., and MCKENZIE, D. (1988), *The Relationship between Plate Motions and Seismic Moment Tensors, and the Rates of Active Deformation in the Mediterranean and Middle East*, Geophys. J. *93*, 45–73.

IOANNIDOU, E. (1989), *Seismic Source Parameters Using Body Wave Inversion: Greece and the Surrounding Area*, Ph.D. Thesis, University of Athens.

KIRATZI, A. (1990), *Moment Tensor Inversion of Some Large Earthquakes in Western Turkey* (in preparation).

KIRATZI, A., and PAPAZACHOS, B. (1984), *Magnitude Scales for Earthquakes in Greece*, Bull. Seismol. Soc. Am. *74*, 969–986.

KIRATZI, A., KARAKAISIS, G., PAPADIMITRIOU, E., and PAPAZACHOS, B. (1985), *Seismic Source — Parameter Relations for Earthquakes in Greece*, Pure Appl. Geophys. *123*, 27–41.

KIRATZI, A., WAGNER, G., and LANGSTON, C. (1991), *Source Parameters of Some Large Earthquakes in Northern Aegean Determined by Body Waveform Inversion*, Pure Appl. Geophys. *135*, 515–527.

KOSTROV, V. (1974), *Seismic Moment and Energy of Earthquakes, and Seismic Flow of Rock*, Izv. Acad. Sci. USSR Phys. Solid Earth *1*, 23–44.

LYBERIS, N. (1981), *Tectonic Evolution of the North Aegean Trough*, J. Struct. Geol. *3*, 709–725.

MCKENZIE, D. (1972), *Active Tectonics of the Mediterranean Region*, Geophys. J. R. Astr. Soc. *30*, 109–185.

MCKENZIE, D. (1978), *Active Tectonics of the Alpine-Himalayan Belt: The Aegean Sea and Surrounding Regions*, Geophys. J. R. Astr. Soc. *55*, 217–254.

MOLNAR, P., and DENG, Q. (1984), *Faulting Associated with Large Earthquakes and the Average Rate of Deformation in Central and Eastern Asia*, J. Geophys. Res. *89*, 6203–6228.

PAPAZACHOS, B. (1989). *Measures of Seismic Strength in the Aegean and Surrounding Area*, 1st Greek Conference on Geophysics, 19–21 April 1989, Athens, 10 pp.

PAPAZACHOS, B. (1990), *Seismicity of the Aegean and Surrounding Area*, Tectonophysics *178*, 287–308.

PAPAZACHOS, B., and PAPAZACHOS, C., *The Earthquakes of Greece* (Ziti Publications, Thessaloniki 1989) 356 pp.

PAPAZACHOS, B., KIRATZI, A., VOIDOMATIS, Ph., and PAPAIOANNOU, C. (1984), *A Study of the December 1981–January 1982 Seismic Activity in Northern Aegean Sea*, Boll. Geofis. Teor. Appl. *26*, 101–113.

ROCCA, A., KARAKAISIS, G., KARACOSTAS, B., KIRATZI, A., SCORDILIS, E., and PAPAZACHOS, B. (1985), *Further Evidence on the Strike-slip Faulting of the Northern Aegean Trough Based on Properties of the August–November 1983 Seismic Sequence*, Boll. Geof. Teor. Applic. *106*, 101–109.

(Received February 11, 1991, revised/accepted May 21, 1991)

PAGEOPH, Vol. 136, No. 4 (1991)

0033–4553/91/040433–16$1.50 + 0.20/0

Regional Stresses Along the Eurasia-Africa Plate Boundary Derived from Focal Mechanisms of Large Earthquakes

Agustín Udías and Elisa Buforn[1]

Abstract —The focal mechanism solutions of 83 European earthquakes with $M > 6$, selected from a total of 140, have been used to derive the directions of the principal axes of stress along the plate boundary between Eurasia and Africa from the Azores islands to the Caucasus mountains. Along most of the region, the horizontal P-axes are at an angle of 45° to 90° with the trend of the plate boundary. Horizontal T-axes are concentrated in central Italy and northern Greece in association with normal faulting. Large strike-slip motion of right-lateral character takes place at the center of the Azores-Gibraltar fault and the North Anatolian fault. From Gibraltar to the Caucasus the boundary is complicated by the presence of secondary blocks and zones of extended deformations with earthquakes spread over wide areas. Intermediate and deep earthquakes are present at four areas with arc-like structure, namely, Gibraltar, Sicily-Calabria, Hellenic arc and Carpathians.

Key words: Eurasia–Africa plate boundary, Mediterranean region, regional stresses, seismicity, focal mechanism, seismotectonics.

Introduction

The boundary between the lithospheric plates of Eurasia and Africa extends from west to east from the Azores islands to the Caucasus mountains where the African plate limits with the Arabia plate. From the Azores to Gibraltar the boundary is relatively simple separating oceanic lithosphere on both sides. East of Gibraltar the boundary is formed by the interaction of continental lithosphere and the oceanic parts of the Mediterranean Sea. The boundary in this region is especially complicated by the presence of small lithospheric blocks and the distribution of stresses by deformations extended over wide areas. The plate boundary must be interpreted in this region as an extended area that follows a complicated system of continental blocks, oceanic basins and orogenic belts, located between the stable parts of Europe and Africa. Earthquakes are also extended over wide areas at the boundary region and intraplate activity is also present.

Many studies have been made concerning the seismotectonic conditions of this region derived from seismicity and focal mechanisms, among them, those of

---

[1] Department of Geophysics, Universidad Complutense de Madrid, Madrid, Spain.

CONSTANTINESCU *et al.* (1966), ISACKS *et al.* (1968), RITSEMA (1969), SHIROKOVA (1972), MCKENZIE (1972), UDÍAS and LOPEZ-ARROYO (1972), PAPAZACHOS (1973), PAYO (1975), UDÍAS *et al.* (1976), MCKENZIE (1978), UDÍAS (1980, 1982), JACKSON and MCKENZIE (1984), UDÍAS (1985), GRIMISON and CHEN (1986), ANDERSON and JACKSON (1987), BUFORN *et al.* (1988), JACKSON and MCKENZIE (1988), PAPAZACHOS (1988), ARGUS *et al.* (1989), WESTAWAY (1990). In this work we study the direction of the stress axes in the region as derived from the focal mechanisms of large earthquakes ($M > 6$) from 1935 to 1983. The deduction of stress directions from fault plane solutions of earthquakes is not exempt from ambiguity. As was pointed out by MCKENZIE (1969) most earthquakes happen on preexisting fault planes of weakness, and slips can occur at different angles relative to the principal axes. In fact, the maximum compressive stress may have an orientation anywhere within the dilatational quadrant, and not necessarily at 45 degrees of the fault plane. This ambiguity may be resolved using mechanisms of many earthquakes in the same area (ANGELIER, 1979; GEPHART and FORSYTH, 1984; RIVERA and CISTERNAS, 1990). However, the stress axes derived from fault plane solutions of large earthquakes may serve as an indication of their general trend for a given region. Using only large earthquakes is advantageous in that their solutions are well determined and the directions of the stresses derived from them correspond more likely to the regional stresses, since the ruptures extend along many kilometers and may not deviate strongly from the expected direction. The consistency in the stress directions found in this study confirms this point of view.

Seismicity

The distribution of epicenters for shallow earthquakes ($h \leq 60$ km) with magnitude $M \geq 4$ from 1960 to 1983 is shown in Figure 1. Earthquakes are located in a general West-East trend occupying a wide band from 30°N to 50°N. Inside this band the trend of the epicenters changes direction several times, outlining secondary blocks. The main characteristics of the seismicity may be described as follows. At the western end of the plate boundary, earthquakes are located following the trend of the Azores islands, branching from the Mid-Atlantic ridge in SE direction. From this point (24°W) earthquakes are located along a West-East transform fault, the Azores-Gibraltar fault, where large earthquakes ($M \geq 8$) are relatively frequent. From about 12°W, epicenters are spread over a wider zone in south Iberia and northern Morocco. In northern Iberia shocks are located along the Pyrenees. From northern Morocco earthquakes continue eastward along the coast of Algeria and Tunisia. From this point, epicenters change direction to the NE in the Sicily-Calabria arc and then continue in NW direction along the Apennines in the Italian peninsula. In northern Italy earthquakes form a wide arc at the Alps and continue again in SE direction along the coast of Yugoslavia and northern Greece. From

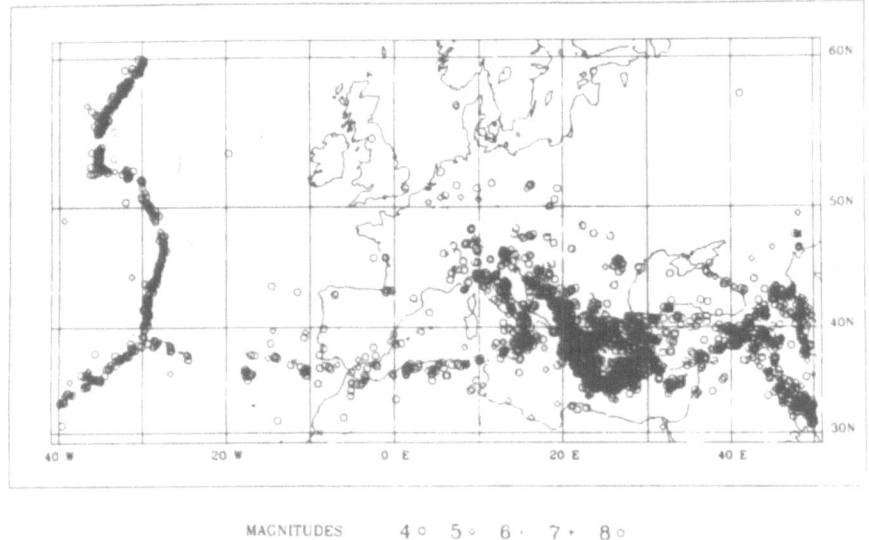

MAGNITUDES   4 ○ 5 ◦ 6 · 7 · 8 ○

Figure 1

Epicenter distribution for shallow earthquakes ($h < 60$ km) and $M > 4$, 1962–1985 (USGS Hypocenter Data File).

here the earthquakes form a wide arc, the Hellenic arc, with the highest intensity of seismic activity in the whole region. The arc in convex to the south and behind it in northern Greece and western Turkey another region of high activity is present with E-W trend. To the east of the Hellenic arc, the arc of Cyprus, a smaller and less active arc, is located. Two alignments of earthquakes are located in the Anatolia peninsula. One at the north in E-W direction along the north Anatolian fault and the other in SW-NE. Both merge at about 40°E where earthquakes continue W-E in the Caucasus and NW-SE along the border of the Arabian plate.

The location of intermediate and deep earthquakes ($h > 60$ km) for the period 1910–1979 and $M > 4$, is shown in Figure 2. They are located at four distinct areas and related to the arc-like structures of Gibraltar, Sicily-Calabria, Hellenic and Carpathian arcs. Deep activity associated with the Gibraltar arc is revealed mainly by the deep Spanish earthquake of 1954 with $M = 7$ and depth 640 km. Two smaller shocks have occurred at the same depth. Some shocks of intermediate depth ($60 < h < 150$ km) are also present in south Spain and north Morocco. Deep earthquakes in the Sicily-Calabria arc are located at its concave side and extend to a depth of 450 km. Most are concentrated between 200 and 350 km with only a few shocks at greater depth. The distribution of shocks and their depth define a subduction zone that trends roughly N-S and dips about 60° in a westerly direction under the Tyrrhenian Sea. A gap of seismic activity is present between 60 and 200 km depth. Deep seismic activity associated with the Carpathian arc is located in a rather small region, known as the Vrancea seismic zone. Most shocks are located

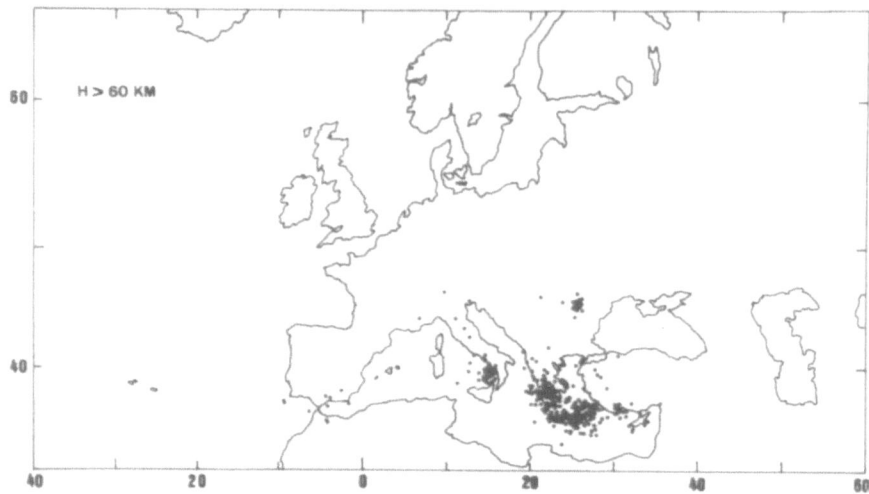

Figure 2
Epicenter distribution for intermediate and deep earthquakes ($h > 60$ km) and $M > 4$, 1910–1985
(USGS Hypocenter Data File).

at depths between 70 and 160 km. Distribution of shocks with depth suggests a nearly vertical subduction zone with earthquakes concentrated within a small volume. The largest concentration of intermediate and deep earthquakes is located along the Hellenic arc, which spans an area from the western coast of Greece to the southern coast of Turkey. The distribution of shocks and their depth indicate a well developed Benioff zone that dips from the convex side of the arc, reaching a maximum depth of 200 km. A second smaller arc associated with intermediate earthquakes is located near the island of Cyprus.

Regional Stresses

Recently, UDÍAS et al. (1989) have published a catalogue of focal mechanisms of European earthquakes with magnitudes $M > 6$ for the period 1906 to 1985, including solutions for 140 earthquakes. From these data a selection has been made of 83 earthquakes for which the mechanism is thought to be well determined and reliable. In order to make the selection, solutions in the catalogue have been classified according to classes A, B and C. Earthquakes belonging to class A have several solutions, all of them in agreement. Earthquakes belonging to class B have several solutions with the majority in agreement or only one solution determined by more than one type of data or with a very large number of data ($N > 100$). Finally, to class C are assigned earthquakes with only one solution, based on only one type of data or several solutions that do not agree with each other. In order to avoid dispersion in the data from the poorly determined solutions, only those of class A and a selection of class B have been selected. The hypocentral parameters of the 83 selected shocks are given in Table 1. For each event one solution has been selected. The *P* and *T* axes, quality of the solutions and references are given in Table 2.

In order to study the orientations of the horizontal components of the principal stresses, we represent those for *P* and *T* axes in separate figures. In both cases only those dipping less than 45° are represented. The directions of the horizontal projections of the pressure axes for shallow earthquakes ($h < 60$ km) are shown in Figure 3. In the same figure the trend of the simplified plate boundary, as derived from the seismicity trends, is given (MCKENZIE 1972: UDÍAS, 1982; JACKSON and MCKENZIE, 1988; WESTAWAY, 1990). The first conclusion from Figure 3 is that the data are very consistent and the majority of the horizontal projections of the *P* axes are nearly normal to the plate boundary. In particular from Azores to Tunisia *P* axes form an angle from 60° to 90° in a consistent NNW-SSE direction. They correspond to strike-slip and thrust mechanisms. Along the coast of Yugoslavia and northern Greece *P* axes are normal to the trend of the coast in NEE-SWW direction. In the Hellenic arc, data are more scattered, but a direction normal to the arc is common. Behind the arc, between Greece and Turkey, *P* axes have E-W direction. Along the north Anatolian fault the direction is NW-SE corresponding to strike-slip mechanisms. In the Caucasus region a nearly N-S trend is found normal to the boundary of the Arabian plate corresponding to thrust mechanisms.

Horizontal projections of the tension axes for shallow earthquakes are shown in Figure 4. Since only those dipping less than 45° are given, the axes represented

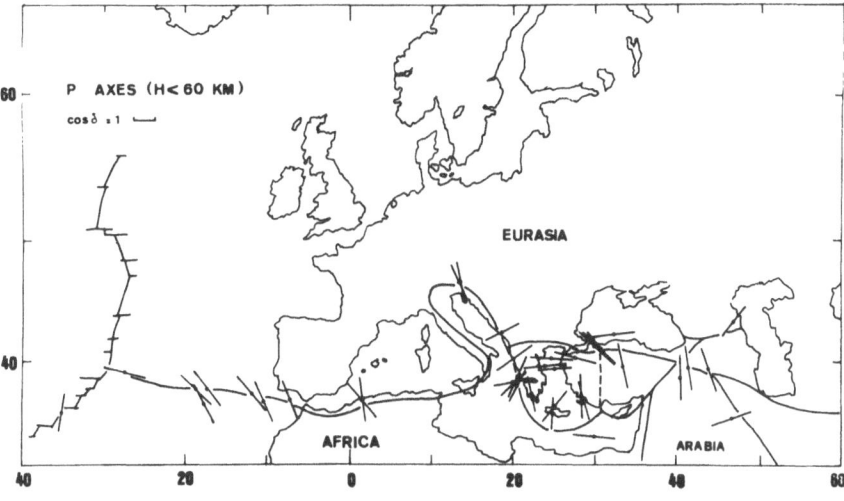

Figure 3
Horizontal projection of *P* axes with plunge less than 45° for shallow earthquakes ($h < 60$ km) and $M > 6$.

Table 1

Hypocenters Parameters

No.	Date	Time	Lat-N	Long-E	Depth	M/M_s
1	1935–02–25	02 51 37	35.75	25.00	80	6.75
2	1938–04–13	02 45 46	39.20	15.20	270	6.75
3	1938–04–19	10 59 15	39.50	33.50		6.75
4	1939–05–08	01 46 50	37.00	−24.50		7.10
5	1939–09–22	00 36 32	39.00	27.00		6.50
6	1940–02–29	16 07 42	35.50	25.50		6
7	1940–10–22	06 37 00	45.75	26.50	150	6.50
8	1940–11–10	01 39 09	45.75	26.50	130	7.40
9	1941–11–25	18 03 55	37.50	−18.50	25	8.40
10	1943–06–20	15 32 53	41.00	30.00		6.25
11	1945–09–02	11 53 57	33.75	28.50	80	6.5
12	1945–09–07	15 48 22	46.00	26.75	100	6.50
13	1946–04–05	20 54 05	35.25	23.50	100	6
14	1947–06–04	00 29 55	40.00	24.00	80	6.00
15	1947–10–06	19 55 37	37.00	22.00		7
16	1948–06–30	12 21 11	38.50	20.50		6.4
17	1949–07–23	15 03 30	38.50	26.50		6.75
18	1951–08–13	18 33 30	40.80	33.40		6.70
19	1953–03–18	19 06 11	40.00	27.50		7.25
20	1954–03–29	06 17 05	37.00	−03.50	640	7.00
21	1954–04–30	13 02 36	39.30	22.20		6.87
22	1954–09–09	01 04 37	36.20	01.60		6.75
23	1955–07–16	07 07 08	37.50	27.00		6.75
24	1955–09–12	06 09 20	32.50	30.00		6.50
25	1956–02–01	15 10 46	39.50	16.00	200	6.20
26	1957–04–25	02 25 42	36.47	28.56	53	7.10
27	1957–05–26	06 33 34	40.67	30.86		7.10
28	1957–05–27	11 01 26	40.50	31.00		6.25
29	1957–12–13	01 45 05	34.41	46.67	42	7.25
30	1959–05–14	06 36 56	35.14	24.58		6.50
31	1959–11–15	17 08 43	37.83	20.47		6.6
32	1960–01–03	20 19 30	39.50	15.50	250	6.20
33	1960–05–26	05 10 05	40.00	20.00		6.30
34	1961–05–23	02 45 16	36.60	28.30	49	6.25
35	1962–08–21	18 19 33	41.40	15.50	34	6.00
36	1962–08–28	10 59 56	37.82	22.89	100	6.70
37	1963–09–18	16 58 13	40.90	29.20	33	6.38
38	1964–03–15	22 30 26	36.20	−07.60	27	6.88
39	1964–10–06	14 31 19	40.30	28.20	10	
40	1965–03–09	17 57 54	39.40	24.00	18	6.38
41	1965–03–31	09 47 31	38.60	22.40	78	6.75
42	1965–07–06	03 18 43	38.40	22.40	20	6.25
43	1966–02–05	02 01 46	39.10	21.70	22	6.25
44	1966–08–19	12 22 11	39.20	41.60	33	6.70
45	1967–01–04	05 58 54	38.44	22.01	10	6
46	1967–03–04	17 58 06	39.22	24.62	33	6.88
47	1967–07–22	16 56 33	40.70	30.80	4	
48	1967–11–30	07 23 52	41.50	20.50	29	6.60

Table 1 (*Contd*)

No.	Date	Time	Lat-N	Long-E	Depth	M/M_s
49	1968–09–03	08 19 52	41.79	32.31	5	6.60
50	1969–02–28	02 40 33	36.01	−10.57	22	8.00
51	1969–03–28	01 48 30	38.59	28.45	9	6.40
52	1969–09–06	14 30 40	36.94	−11.89	33	6.00
53	1970–03–28	21 02 23	39.18	29.49	20	7.10
54	1970–04–08	13 50 27	38.43	22.66	17	5.90
55	1970–05–14	18 12 28	43.03	47.09	44	6.50
56	1970–11–18	12 23 18	35.15	−35.74	33	6
57	1971–05–12	06 25 13	37.59	29.76	23	6.3
58	1972–05–04	21 40 09	35.12	23.61	46	6.30
59	1972–09–13	04 13 21	37.93	22.39	83	
60	1972–09–17	14 07 16	38.28	20.34	33	6.3
61	1975–05–26	09 11 51	35.99	−17.65	33	7.9
62	1975–09–06	09 20 11	38.47	40.72	26	6.70
63	1976–05–06	20 00 12	46.36	13.28	9	6.5
64	1976–07–28	20 17 42	43.17	45.60	21	6.10
65	1976–09–15	03 15 20	46.30	13.20	10	6
66	1976–11–24	12 22 19	39.12	44.03	36	7.30
67	1977–03–04	19 21 54	45.77	26.76	94	7.20
68	1978–06–20	20 03 21	40.74	23.23	3	6.40
69	1978–11–04	15 22 20	37.71	48.95	34	6.1
70	1979–05–24	17 23 18	42.24	18.75	5	6.4
71	1980–01–01	16 42 40	38.80	−27.80		6.70
72	1980–07–09	02 35 52	39.23	22.59	20	6.1
73	1980–10–10	12 25 24	36.20	01.30		7.30
74	1980–11–23	18 34 54	40.90	15.40		6.90
75	1981–02–24	20 53 37	38.22	22.97	18	6.6
76	1981–02–25	02 35 54	38.17	23.12	30	6.3
77	1981–03–04	21 58 07	38.24	23.26	32	6.4
78	1981–12–19	14 10 51	39.22	25.25	10	7.2
79	1981–12–27	17 39 13	38.91	24.92	10	6.5
80	1983–01–17	12 41 30	38.07	20.24	14	7
81	1983–03–23	23 51 05	38.23	20.29	13	6.2
82	1983–08–06	15 43 52	40.14	24.74	2	6.9
83	1983–10–17	19 36 22	37.59	−17.41	10	6.3

correspond to the regions under horizontal tensional regime. In the Azores T axes are normal to the volcanic alignment. Along the Azores-Gibraltar fault, horizontal T axes in NE-SW correspond to the horizontal P axes, resulting in strike-slip right-lateral motion along the fault. Horizontal T axes are present in the Apennines, Italy in NE-SW direction, normal to the trend of the mountain chain and corresponding to normal faulting. A large concentration of horizontal T axes in a general N-S direction is present behind the Hellenic arc, in Greece and western Turkey. Along the north Anatolian fault, horizontal T axes in NE-SW direction correspond to the also horizontal P axes in agreement with the strike-slip motion along the fault.

A. Udías and E. Buforn

Table 2

Fault-Plane Solutions

No.	P		T		Q	Ref.
	Trend	Plunge	Trend	Plunge		
1	344	21	212	60	B	18
2	280	35	80	55	B	19
3	348	14	254	16	A	11
4	234	49	355	24	B	3
5	70	74	176	05	B	18
6	220	30	40	60	B	18
7	130	10	162	75	A	10
8	303	11	140	78	A	18
9	133	11	42	04	A	3
10	132	10	40	10	A	15
11	130	00	00	90	B	18
12	290	10	182	59	A	18
13	249	31	69	59	B	18
14	232	25	88	59	B	18
15	168	35	258	01	B	20
16	205	04	301	57	B	18
17	102	45	197	05	A	15
18	302	21	37	11	A	15
19	284	02	15	11	A	15
20	87	43	279	46	A	9
21	51	72	253	17	A	18
22	171	15	11	74	A	15
23	261	57	354	02	A	15
24	276	05	180	52	B	15
25	324	62	118	25	A	4
26	163	25	262	19	A	18
27	312	05	43	12	A	15
28	329	14	234	18	B	20
29	71	01	340	62	A	15
30	182	44	76	16	A	18
31	22	37	235	50	A	1
32	278	59	95	32	A	7
33	57	16	324	12	A	18
34	179	17	359	73	A	15
35	331	81	205	06	A	15
36	179	01	87	68	A	15
37	134	57	23	17	A	11
38	158	25	304	61	A	3
39	354	80	174	10	A	20
40	265	00	175	03	B	15
41	235	27	80	60	A	15
42	358	59	177	31	A	15
43	143	67	350	20	B	1
44	353	04	259	42	A	20
45	170	45	350	45	B	20
46	328	82	212	04	A	15
47	138	06	228	01	A	20

Table 2 (*Contd*)

No.	P Trend	Plunge	T Trend	Plunge	Q	Ref.
49	84	01	174	36	A	11
50	165	04	67	64	A	3
51	6	96	186	21	A	18
52	138	14	47	07	A	3
53	84	61	200	14	A	18
54	346	64	187	25	A	14
55	40	29	225	61	B	11
56	190	18	283	01	A	3
57	34	78	144	04	B	19
58	196	49	16	41	A	14
59	355	10	92	56	A	14
60	260	29	258	14	B	1
61	151	09	246	10	A	3
62	359	01	91	62	A	11
63	168	30	342	60	B	1
64	354	34	174	56	B	11
65	155	10	268	66	B	1
66	340	12	71	11	A	11
67	323	17	151	73	B	8
68	315	83	177	15	A	2
69	90	25	270	65	B	11
70	244	25	64	65	A	1
71	104	15	194	02	A	3
72	112	82	344	02	A	17
73	320	09	106	80	A	6
74	183	68	37	18	A	5
75	298	76	181	07	A	21
76	270	83	153	03	A	21
77	33	86	154	02	A	21
78	255	35	357	16	A	17
79	78	06	168	02	A	17
80	225	38	44	52	A	1
81	249	19	348	25	A	1
82	272	04	02	05	A	16
83	320	04	50	01	A	3

References:
1. ANDERSON and JACKSON (1987)
2. BUFORN (1982)
3. BUFORN et al. (1988)
4. CANITEZ and UCER (1967)
5. CELLO et al. (1982)
6. DESCHAMPS et al. (1982)
7. GASPARINI et al. (1983)
8. GIRARDINI et al. (1984)
9. HODGSON and COCK (1956)
10. ISACKS and MOLNAR (1971)
11. JACKSON and McKENZIE (1984)
12. KANAMORI and GIVEN (1981)

14. McKENZIE (1978)
15. McKENZIE (1972)
16. PAPAZACHOS et al. (1984a)
17. PAPAZACHOS et al. (1984b)
18. RITSEMA (1974)
19. RIUSCETTI and SCHICK (1975)
20. STEWART and KANAMORI (1982)
21. WON-YOUNG KIM et al. (1984).

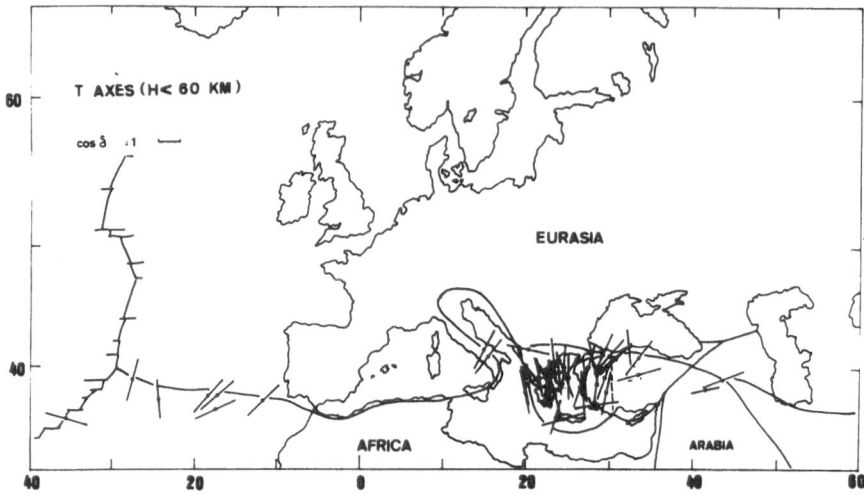

Figure 4
Horizontal projection of T axes with plunge less than 45° for shallow earthquakes ($h < 60$ km) and
$M > 6$.

For intermediate and deep shocks, directions of the horizontal projections of P
axes are shown in Figure 5. In south Spain, the very deep earthquake (640 km) has
the P axis in E-W direction. In Sicily-Calabria, P axes are normal to the trend of
the arc and the same is the case in the Carpathian arc. Intermediate shocks in the
Hellenic arc have a greater dispersion. Two trends nearly perpendicular in NNW-
SEE and NE-SW directions are present. Vertical projections of the P axes for each

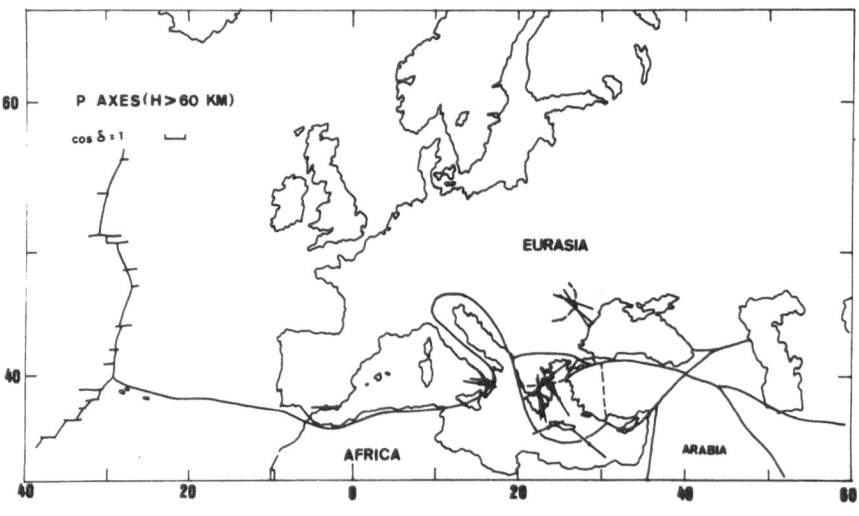

Figure 5
Horizontal projection of P axes for intermediate and deep earthquakes ($h > 60$ km) and $M > 6$.

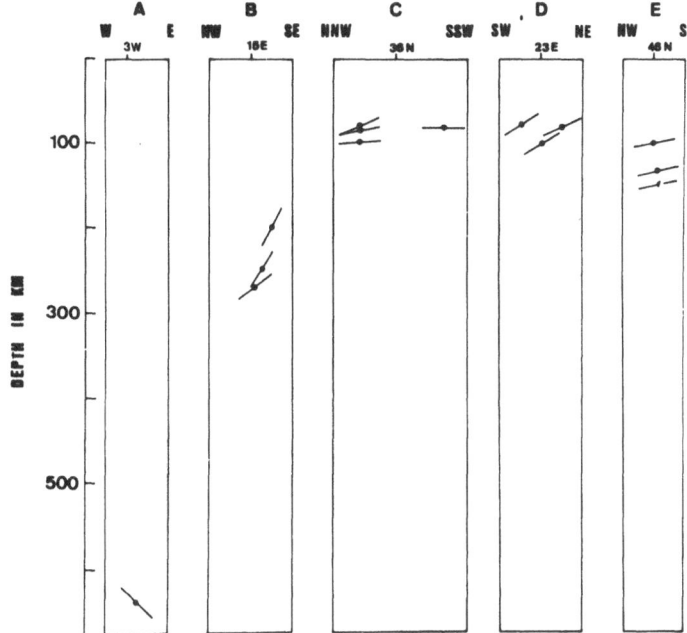

Figure 6

Projection on a vertical plane, of the *P* axes for intermediate and deep earthquakes, containing their average direction for the regions: A—Gibraltar, B—Sicily-Calabria, C and D—Hellenic arc, and E—Carpathians.

of the four regions are given in Figure 6. In this figure the axes are projected on a vertical plane containing the average direction of the axes. For the Hellenic arc, two planes are used, one in NNW-SSW and the other in SW-NE direction, according to the two sets of orientations found. The deep Spanish earthquake (A) is the only one with the axis dipping 45° to the East, all the other dip to the NW or SW. In Sicily-Calabria (B) the shocks at depths between 200 and 300 dip steeply to the NW. In the Carpathian region (E) the *P* axes are nearly horizontal, dipping slightly to the NW. In the Hellenic arc (C and D) the shocks are at about 100 km depth and their *P* axes are nearly horizontal. Those of set D with dips to the SW have somewhat larger dips.

Slip Vectors

In Figure 7 the directions of the slip vectors for the shallow shocks are shown corresponding to the African plate. In order to select the fault plane, we have assumed that the motion in the strike-slip faults is in the E-W direction and in the dip-slip faults, corresponding to down going of Africa with respect to Eurasia.

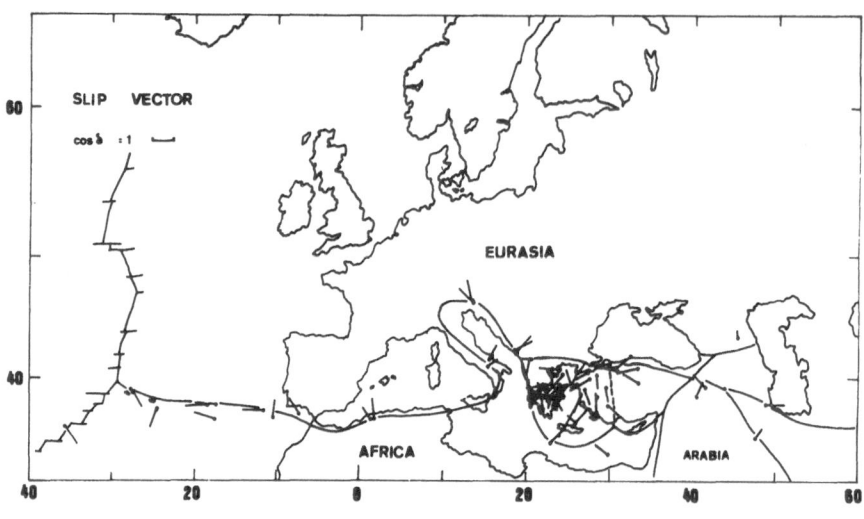

Figure 7
Horizontal projection of the slip vectors corresponding to the Africa plate.

From Azores to Tunisia, the slip vector rotates from south to west and finally to north. These directions of the slip vectors are consistent with a location of the pole of counterclock rotation of the Africa plate with respect to Eurasia plate near the Canary islands (21°N, 20°W) (ARGUS *et al.*, 1989). In central Italy (Adriatic block) the slip is directed NE, the same direction found for the shocks in the coast of Yugoslavia. In north Italy, however, the slips are in NNW direction. This situation is the main argument (ANDERSON and JACKSON, 1987) for the independent motion of the Adriatic block about a different pole of counterclock rotation at about 46°N, 10°E with respect to Eurasia.

Along the Hellenic arc, Africa moves to the north in a way consistent with the subduction process. Behind the arc, slip vectors on the Aegean block, including the western part of the Anatolian peninsula, have a consistent trend to the south. In the Anatolian block, however, the slip vectors point to the west. These two blocks must move, then, differentially with respect to Europe and Africa. Finally, the slip vectors on the Arabia plate all have a northerly trend.

Conclusions

Figure 8 shows a simplified summary picture of the situation along the plate boundary, based on the stress and slip directions presented in this work. The broad line of the figure follows the alignment of the most active seismicity (Fig. 1) and represents, in a broad sense, the location of the plate boundary. This line, however,

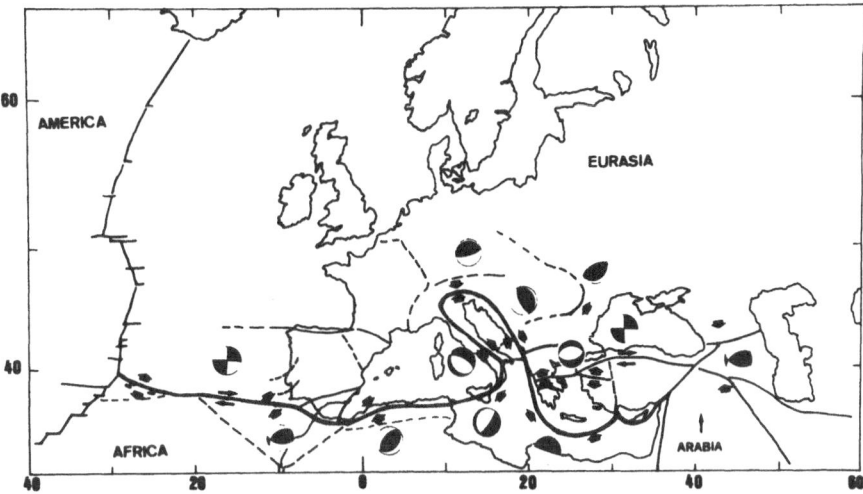

Figure 8
Seismotectonics of the plate boundary between Eurasia and Africa showing the direction of the regional
stresses and predominant focal mechanism.

must be understood as a simplification of the actual situation that involves broad
areas of deformation and cannot be reduced totally to the motion of rigid blocks.
Thinner and dashed lines represent the limits of areas that form secondary small
blocks which have independent or semi-independent movement with respect to the
two main plates, or areas of deformations, and are based on the distribution of
seismicity. The predominant type of focal mechanism in each region, the directions
of the horizontal stresses and of the horizontal slip are also shown in Figure 8.

From West to East, the main characteristics of the plate boundary region are
the following: At the Azores triple junction, the boundary between Africa and
Eurasia has a ridge nature and is under horizontal normal tensions. From about
23°W to 12°W, the boundary is formed by a transform fault with strike-slip,
right-lateral motion, with Africa moving West with respect to Eurasia. Near 12°W
the motion changes to reverse faulting with the Africa plate moving under Eurasia.
This type of motion is found in Algeria and may extend to 10°E. These changes at
such a short distance along the boundary are consistent with a pole of rotation for
Africa, not far from the boundary itself at about 21°N, 20°W. The presence of
earthquakes extending into the Atlantic in front of the coast of Portugal, in
Portugal, south Spain and northern Morocco and Algeria points to the presence of
small blocks or deformable areas at both sides of the boundary. The earthquakes in
the Pyrenees and northern Spain suggest that the stable part of the Iberian
peninsula may form a block semi-independent from the Eurasia plate.

In the Sicily-Calabria arc, the lithospheric material of the Africa plate is
subducted, pushed from the SE and reaching about 350 km in depth. In the

subducted plate, the stresses dip steeply to the NW (Fig. 6). The Adriatic block has a motion that, as previously stated, does not agree with the motion of Africa. Along the Apennines, the block moves to the east from the other side of Italy, due to normal faulting. This motion causes reverse faulting and compressional stresses in NE-SW direction along the coast of Yugoslavia. In the northern part, however, the block pushes in N to NW direction producing also reverse faulting. This compression is transmitted to the broad region of the Alps. North of the Alps, a line of deformation extends even further along the Rhine graben. Also inside the Eurasia plate, the seismic activity related to the Carpathian region with a small vertical subducted block with compressional stresses pushing from the SE is located. The region of deformation extends further to the NW. This region may be considered as a former plate or block boundary that is now locked inside the Eurasia plate.

The region related to the Hellenic and Cyprus arcs is quite complex. The arcs themselves are under compressional stresses and lithospheric material of the Africa plate is subducted from the South to depths of about 200 km. Behind the Hellenic arc, the lithosphere is under tensional N-S stresses and predominant normal faulting. This results in stretching of the material and the formation of a back arc basin. This zone of extension covers Greece and western Anatolia. The rest of Anatolia moves westward along the north Anatolian fault, with large horizontal motion of strike-slip rigth-lateral character. This causes the northern part of the back arc basin to be subjected to E-W horizontal pressure (Fig. 3). The Arabia plate moves northward and causes, in one side, the motion to the West of the Anatolia block and the compression of the broad Caucasus region. In this region horizontal compressions have N-S direction and predominant motion takes place on reverse faults.

Acknowledgements

The authors wish to thank J. Ruiz de Gauna for his work in compiling the catalogue of focal mechanisms and Dr. A. Espinosa, USGS, Golden, Col. for providing seismicity data and helpful suggestions. This work has been partially supported by the Dirección General de Investigación Científica y Tecnica, project PB-89-0097 and the European Community, project SCI-0176-C-(SMA). Publication No. 328, Departamento de Geofísica, Universidad Complutense de Madrid.

REFERENCES

ANDERSON, H., and JACKSON, J. (1987), *Active Tectonics of the Adriatic Region*, Geophysics J. R. Astr. Soc. *91*, 937–983.
ANGELIER, J. (1979), *Determination of the Mean Principal Directions of Stresses for a Given Fault Population*, Tectonophysics *56*, 17–26.

ARGUS, D. F., GORDON, R. G., DeMETS, C., and STEIN, S. (1989), *Closure of the Africa-Eurasia-North America Plate Motion Circuit and Tectonics of the Gloria Fault*, J. Geophys. Res. *94*, 5585–5602.

BUFORN, E. (1982), *Estudio estadístico de la dirección de esfuerzos principales en terremotos*, Doctoral Thesis, Universidad Complutense, Madrid, Spain.

BUFORN, E., UDÍAS, A., and COLOMBAS, M. A. (1988). *Seismicity, Source Mechanisms and Tectonics of the Azores-Gibraltar Plate Boundary*, Tectonophysics *151*, 89–118.

CANITEZ, N., and UCER S. (1967), *Computer Determinations of the Fault Plane Solutions in Near Anatolia*, Tectonophysics *4*, 235–244.

CELLO, G., GUERRA, I., TORTORICI, L., TURCO E., and SCARPA, R. (1982), *Geometry of the Neotectonic Stress Field in Southern Italy: Geological and Seismological Evidence*, J. Struct. Geology *4*, 385–393.

CONSTANTINESCU, L., RUPRECHTOVA, L., and ENESCU, D. (1966), *Mediterranean-Alpine Earthquake Mechanisms and their Seismotectonic Implications*, Geophys. J. R. Astr. Soc. *10*, 347–368.

DESCHAMPS, A., GAUDEMER, Y., and CISTERNAS, A. (1982), *The El Asnam, Argelia, Earthquake of 10 October 1980: Multiple Source Mechanism Determined from Long Period*, Bull. Seismol. Soc. Am. *72*, 1111–1128.

GASPARINI, C., and IANNACCONE, G., and SCARPA R. (1983), *Fault Plane Solutions and Seismicity of the Italian Peninsula*, Tectonophysics *117*, 59–78.

GEPHART, J. W., and FORSYTH, D. W. (1984), *An Improved Method for Determining the Regional Stress Tensor Using Earhtquake Focal Mechanism Data: Application to the San Fernando Earthquake Sequence*, J. Geophys. Res. *89*, 9305–9320.

GIARDINI, D., DZIEWONSKI, A. M., WOODHOUSE, J. W., and BOSCHI, E. (1984), *Systematic Analysis of the Seismicity of the Mediterranean Region using the Centroid-moment Tensor Method*, The O.G.S. Silver Anniversary Volume *121*, 141.

GRIMISON, N. L., and CHEN, W. (1986), *The Azores-Gibraltar Plate Boundary: Focal Mechanism, Depths of Earthquakes and their Tectonic Implications*, J. Geophys. Res. *92*, 2029–2047.

HODGSON, J. H., and COCK I. J. (1956), *Direction of Faulting in Spanish Earthquake 1954*, Tellus *8*, 321–327.

ISACKS, B., and MOLNAR, P. (1971), *Distribution of Stresses in the Descending Lithosphere from a Global Survey of Focal Mechanism Solutions of Mantle Earthquakes*, Rev. of Geophys. and Space Phys. *9*, 1–17.

ISACKS, B., OLIVER, J., and SYKES, L. R. (1968), *Seismology and the New Global Tectonics*, J. Geophys. Res. *73*, 5855–5899.

JACKSON, J., and McKENZIE, D. (1984), *Active Tectonics of the Alpine-Himalayan Belt between Western Turkey and Pakistan*, Geophys. J. R. Astr. Soc. *77*, 185–264.

JACKSON, J., and McKENZIE, D. (1988), *The Relationship between Plate Motions and Seismic Moment Tensors and the Rates of Active Deformation in the Mediterranean and Middle East*, Geophys. J. R. Astr. Soc. *93*, 45–73.

KANAMORI, H., and GIVEN, J. W. (1981), *Use of Long-period Surface Waves for Rapid Determination of Earthquake Source Parameters*, Phys. Earth Plan. Int. *27*, 8–31.

McKENZIE, D. (1969), *The Relation between Fault-plane Solutions for Earthquakes and the Directions of the Principal Stresses*, Bull. Seismol. Soc. Am. *59*, 591–601.

McKENZIE, D. (1972), *Active Tectonics of the Mediterranean Region*, Geophys. J. R. Astr. Soc. *30*, 109–185.

McKENZIE, D. (1978), *Active Tectonics of the Alpine-Himalayan Belt: The Aegean Sea and Surrounding Regions*, Geophys. J. R. Astr. Soc. *55*, 217–254.

PAPAZACHOS, B. C. (1973), *Distribution of Seismic Foci in the Mediterranean and Surrounding Area and its Tectonic Implication*, Geophys. J. R. Astr. Soc. *33*, 421–430.

PAPAZACHOS, B. C., COMNINAKIS, E. E., PAPADIMITRIOU, E. E., and SCORDILIS, E. M. (1984a), *Properties of the February-March 1981 Seismic Sequence in the Alkyoinides Gulf of Central Greece*, Ann. Geophys. *2*, 537–544.

PAPAZACHOS, B. C., KIRATZI, A. A., VOIDOMATIS, P., and PAPAIOANNOU, C. (1984b), *A study of the December 1981-January 1982 Seismic Activity in the Northern Aegean Sea*, Boll. Geof. Teor. Appl. *26*, 101–113.

PAPAZACHOS, B. C., *Active tectonics in the Aegean and surrounding area*, In *Seismic Hazard in Mediterranean Regions* (J. Bonin et al., eds.) (Kluger Academic Pub., Dordrecht 1988) pp. 301–331.

PAYO, G. (1975), *Estructura, sismicidad y tectónica del mar Mediterráneo*, Instituto Geográfico y Catastral, Madrid. Monografía, 39 pp.

RIVERA, L. A., and CISTERNAS, A. (1990), *Stress Tensor and Fault Plane Solutions for a Population of Earthquakes*, Bull. Seismol. Soc. Am. *80*, 600–614.

RITSEMA, A. R. (1969), *Seismic Data of the West Mediterranean and the Problem of Oceanization*, Verhand K. Ned. Geol. Mijn. Gen. *26*, 105–120.

RITSEMA, A. R. (1974), *The Earthquake Mechanisms of the Balkan Region*, Koninklijk Nederlands Meteorologisch Instituut, De Bilt.

RIUSCETTI, M., and SCHICK, R. (1975), *Earthquakes and Tectonics in Southern Italy*, Boll. di Geofisica *17*, 59–78.

SHIROKOVA, E. I., *Stress Pattern and Probable Motion in Earthquake Foci of the Asia-Mediterranean Seismic Belt. Field of the Earth's Elastic Stresses and Mechanism of Earthquake Foci*, (Publishing House Nauka, Moscow 8, 1972) pp. 112–148.

STEWART, G. S., and KANAMORI H. (1982), *Complexity of Rupture in Large Strike-slip Earthquakes in Turkey*, Phys. Earth Plan. Int. *28*, 70–84.

UDÍAS, A. (1980), *Seismic stresses in the region Azores-Spain-Western Mediterranean*, In *Tectonic Stresses in the Alpine-Mediterranean Region*, Rock Mech. Suppl. *9*, 75–84.

UDÍAS, A. (1982), *Seismicity and seismotectonics stress field in the Alpine-Mediterranean region*, In *Alpine-Mediterranean Geodynamics* (Berckhemer, H. and Hsü, J. K. eds.) Geodynamics Series *7*, 75–82.

UDÍAS, A., *Seismicity of the Mediterranean basin*, In *Geological Evolution of the Mediterranean Basin* (Stanley, D. J. and Wezel, F. C., eds.) (Springer-Verlag, New York 1985) pp. 55–63.

UDÍAS, A., and LOPEZ ARROYO, A. (1972), *Plate Tectonics and the Azores-Gibraltar Region*, Nature *237*, 67–69.

UDÍAS A., LOPEZ ARROYO, A., and MEZCUA J. (1976), *Seismotectonics of the Azores-Alboram Region*, Tectonophysics *31*, 259–289.

UDÍAS, A., BUFORN, E., and RUIZ de GAUNA, J. (1989), *Catalogue of Focal Mechanisms of European Earthquakes*, Dept. Geophysics, Universidad Complutense, Madrid.

WESTAWAY, R. (1990), *Present-day Kinematics of the Plate Boundary Zone between Africa and Europe, from Azores to the Aegean*, Earth and Plan. Sci. Lett. *96*, 393–406.

WON-YOUNG KIM, KULHANEK, O., and MEYER, K. (1984), *Source Processes of the 1981 Gulf of Corinth Earthquake Sequence from Body-wave Analysis*, Bull. Seismol. Soc. Am. *74*, 459–477.

(Received November 15, 1990, revised/accepted February 25, 1991)

PAGEOPH, Vol. 136, No. 4 (1991)

0033–4553/91/040449–10$1.50 + 0.20/0

Focal Mechanisms of Intraplate Earthquakes in Bolivia, South America

A. VEGA[1] and E. BUFORN[2]

Abstract —Intraplate seismic activity in Bolivia is mainly located in the central region (16°–19°S, 63°–67°W) which includes the East Andean Cordillera and the Sub-Andean Sierras. At this region there is a bend in the trend of the main geological structures from NW-SE in the north to N-S in the south. Focal mechanisms have been calculated for 10 earthquakes of magnitudes 4.9–5.6, using first motion *P*-waves from long period instruments. Their solutions correspond to reverse faulting, some with a large component of strike-slip motion. Their solutions can be grouped into two types; one with pure reverse faulting on planes with azimuth NW-SE and the other with a large strike-slip component on planes with azimuths nearly N-S or WNW-ESE. The maximum stress axis (*P*-axis) is practically horizontal (dipping less than 5°) oriented in a mean N56°E direction. This orientation may be related with the direction of compression resulting from the collision of the Nazca plate against the western margin of the South American continent. Wave-form analysis of long-period *P*-waves for one event restricts the focal depth to 8 km in the Sub-Andean region. Seismic moments and source dimensions determined from spectra of Rayleigh waves are in the range of 10^{16}–10^{17} Nm and 17–24 km, respectively. The Central Bolivia region can be considered as a zone of intraplate deformation situated between the Bolivian Altiplano and the Brazil shield.

Key words: Seismicity, focal mechanism, intraplate earthquakes, Bolivia, Andean Cordillera, seismotectonics, South America.

Introduction

The Bolivian Andean Cordillera, extending from 10°S to 35°S, forms part of the Andean chain in South America. The origin of this chain is a slow subduction process, beginning in the Caenozoic, with lithospheric material in the Pacific Ocean pushed under the western part of the continent (JAMES, 1971). This process has produced large intraplate deformations in a broad region of the continent.

The Bolivian Andean Cordillera may be divided into four zones according to its morphologic, tectonic and paleographic characteristics. From west to east, they are (Figure 1): The Altiplano, with an average altitude of about 4000 m, the East Andean Cordillera, with altitudes higher than 6000 m, the Sub-Andean Sierras, with altitudes lower than 2500 m and the Beni and Chaco Plains, with altitudes lower than 250 m.

[1] Obervatorio San Calixto, La Paz, Bolivia.
[2] Dept. de Geofísica, Universidad Complutense, Madrid, Spain.

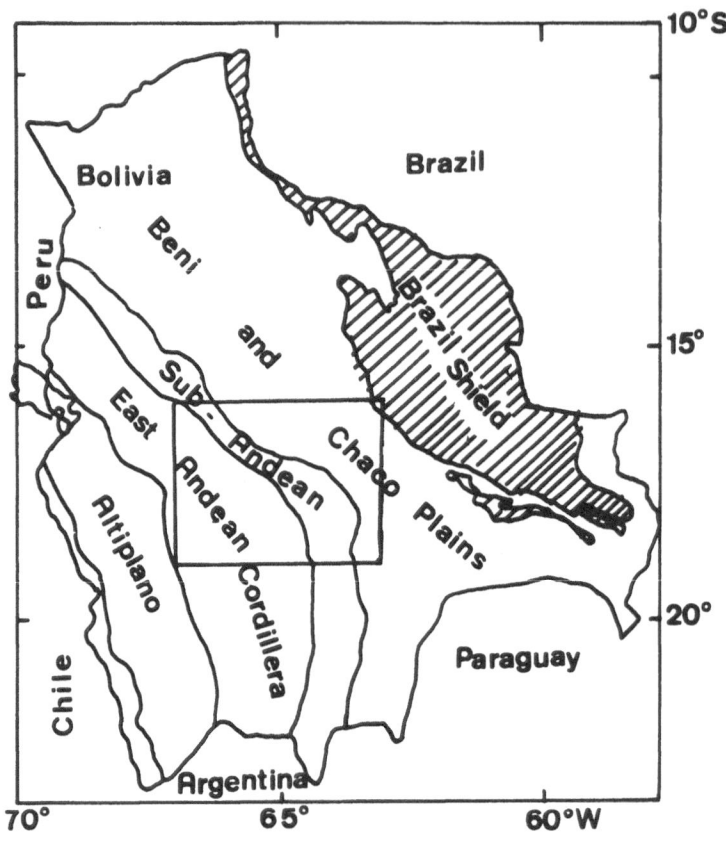

Figure 1
Geological units in Bolivia. Rectangle corresponds to the area studied in this paper.

These geological units are separated by very deep faults and correspond to different terrains: Paleozoic in East Cordillera, Mesozoic in the Altiplano and Sub-Andean Sierras and Quaternary in the Beni and Chaco Plains. Andean Tertiary deformations have affected the Paleozoic and Mesozoic sediments, from the Pacific coast to the eastern limit of the Sub-Andean Sierra, where large faults with NWN-SES trend are present. In Bolivia, the Andean Tertiary deformations are less important than the older Hercynian deformations. Regional Andean metamorphisms are not observed in this region. Extrusive magmatism is present in the Altiplano and intrusive in the East Cordillera, both of Calco-alkaline composition. At 18°S, in the East Cordillera and Sub-Andean Sierra, the general trend of the geological features changes from NW-SE to N-S direction. This change in direction is known as the Santa Cruz "bend".

Seismic activity in Bolivia is moderate compared to the high activity in Peru and Chile (Figure 2). There are, however, some large events at intermediate depth,

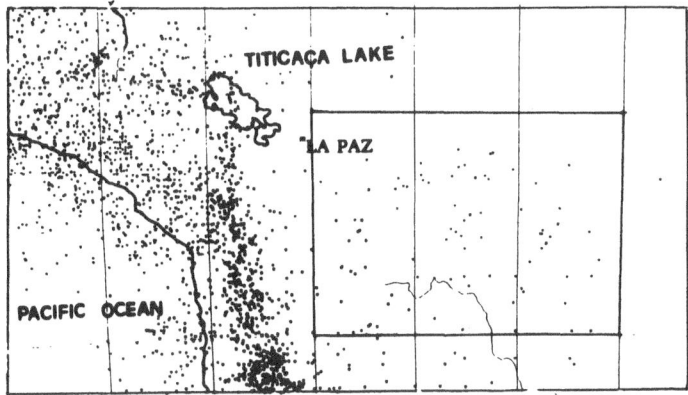

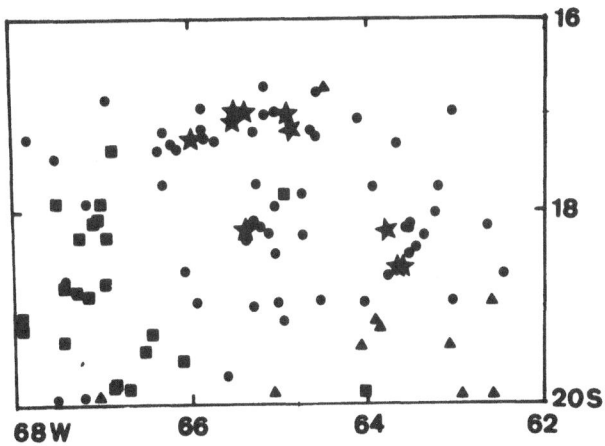

Figure 2

Seismicity of South America corresponding to southern Perú, Bolivia and northern Chile. Rectangle corresponds to the area studied in this paper. Circles correspond to shallow shocks ($h < 60$ km), squares to intermediate shocks ($60 \leq h \leq 300$ km) and triangles to deep earthquakes ($h > 300$ km). Stars show the location of ten events studied in this paper.

between 100 km and 300 km, in the West and SW region near the border with Chile and Peru. Deep shocks ($h > 300$ km) occur in the southern region of Bolivia and northern region of Argentina. In the central region of Bolivia ($16°-19°$S, $63°-67°$W, Figure 2) at the central part of Eastern Andean Cordillera and the Sub-Andean Sierra, the earthquakes are shallow with small magnitude (VEGA, 1985, 1989), with occasional moderate shocks; range magnitude is equal or lower than 5.6, with maximum intensity about VII and foci at shallow depth (DESCOTES and CABRÉ, 1965, 1973; RODRIGUEZ and VEGA, 1976; VEGA, 1976; CABRÉ, 1981; OCOLA, 1984; VEGA, 1986; CABRÉ and VEGA, 1989). Most of the events located in

Bolivia for the period 1913–1981 correspond to shocks occurred in the area studied in this paper (VEGA, 1985).

The seismic activity in Bolivia may be associated in a general way with the subduction of the Nazca plate below South America defining the central region of Bolivia, as a zone of intensive crustal intraplate deformations, even though not very well defined, but producing structural deflections of the Paleozoic blocks forming the Santa Cruz bend.

Focal Mechanisms

Fault-plane solutions for 10 earthquakes located in the studied area (Figure 2), for the period 1968–1986 with magnitudes between 4.9 and 5.6, have been determined. Their focal depth is not very well constrained, but most shocks are considered to be of shallow foci ($h \leq 40$ km) and only one has a deeper focus at 72 km (Table 1). Data used are first motions of P-waves recorded at long period WWSSN stations at teleseismic distances. The solutions obtained for these events, using the algorithm of BRILLINGER et al. (1980), are shown in Table 2 and Figure 3. Events 2, 6, 7 and 10 have very similar solutions corresponding to reverse faults with a plane oriented in NWN-SES direction and horizontal pressure axis in NW-SE direction ($\Phi = 56°$). The solutions for shocks 1, 3, 4, 5, 8 and 9 correspond to reverse faults with a large component of strike-slip motion and horizontal pressure axis oriented also in NW-SE direction. Of the two nodal planes one is oriented nearly in N-S direction with right-lateral motion and other in WNW-ESE direction with left-lateral motion.

From spectral analysis of Rayleigh waves recorded at long period teleseismic WWSSN stations ($\Delta \geq 20°$) scalar seismic moments and dimensions of 9 events have been determined (Table 3). Ben-Menahen model (BEN-MENAHEN, 1970) was

Table 1

Hypocentral coordinates of events studied in this paper

N	Date	Time	Lat.	Long.	Depth (km)	Magn.
1	22–08–1968	00–06–42.	17.2S	64.8W	72	4.9
2	18–07–1969	23–17–09	18.3S	63.3W	8	5.5
3	12–05–1972	17–16–20	17.3S	66.2W	43	5.0
4	22–02–1976	08–09–23	18.3S	65.3W	41	5.2
5	23–07–1981	13–51–26	17.0S	64.9W	38	5.3
6	19–03–1985	10–28–46	18.6S	63.7W	42	5.5
7	22–03–1985	14–02–41	18.6S	63.6W	39	5.5
8	09–05–1986	16–23–59	17.1S	65.6W	13	5.6
9	19–06–1986	20–33–17	17.0S	65.4W	18	5.3
10	19–06–1986	21–57–33	17.0S	65.6W	15	5.4

Table 2

Fault-plane solutions

No.	Axes T, P Φ	θ	Fault Plane ϕ	δ	λ
1	T:326 ± 2	60 ± 5	A:105 ± 4	69 ± 7	158 ± 4
	P: 56 ± 5	89 ± 6	B: 7 ± 3	70 ± 4	22 ± 3
2	T:318 ± 34	19 ± 17	A:139 ± 17	41 ± 3	117 ± 24
	P: 69 ± 7	83 ± 7	B:354 ± 36	54 ± 10	112 ± 16
3	T:307 ± 13	46 ± 12	A: 97 ± 13	53 ± 10	152 ± 7
	P: 46 ± 13	81 ± 10	B:349 ± 13	68 ± 6	140 ± 11
4	T:334 ± 11	50 ± 12	A:117 ± 18	61 ± 23	151 ± 15
	P: 66 ± 22	87 ± 23	B: 12 ± 15	64 ± 17	33 ± 23
5	T:323 ± 12	61 ± 11	A:103 ± 11	66 ± 21	163 ± 11
	P: 56 ± 10	84 ± 12	B: 6 ± 12	75 ± 9	25 ± 22
6	T:291 ± 61	11 ± 9	A:136 ± 12	39 ± 7	104 ± 18
	P: 56 ± 12	84 ± 6	B:334 ± 20	52 ± 5	101 ± 16
7	T:335 ± 10	19 ± 7	A:347 ± 10	46 ± 2	117 ± 10
	P:238 ± 6	88 ± 3	B:131 ± 8	50 ± 5	115 ± 8
8	T:324 ± 5	56 ± 4	A:106 ± 5	64 ± 5	157 ± 2
	P: 56 ± 5	86 ± 4	B: 5 ± 4	70 ± 2	28 ± 5
9	T:316 ± 4	50 ± 8	A:103 ± 7	56 ± 10	155 ± 6
	P: 53 ± 7	82 ± 11	B:358 ± 5	69 ± 4	144 ± 11
10	T:299 ± 23	22 ± 12	A:110 ± 14	44 ± 4	121 ± 17
	P: 41 ± 16	85 ± 9	B:330 ± 21	53 ± 9	117 ± 11

used to estimate the seismic moment. An estimate of the source dimensions has been calculated from the corner frequency of the spectrum of Rayleigh waves f_c using the expression:

$$R = 0.273 * V_R / f_c$$

where V_R is the phase velocity of the Rayleigh waves.

Values of seismic moments for the events 6, 7 and 10 are smaller than those given by NEIC for the same events, by a factor of 1–0.5.

Table 3

Seismic moment and source dimensions

Event	m_b	M_s	$M_0(\text{Nm})$	$2r(\text{km})$	N
18–07–1969	5.5		4.3×10^{16}	19	12
12–05–1972	5.0		1.9×10^{16}	21	2
22–02–1976	5.2		2.3×10^{16}	18	5
23–07–1981	5.0	4.1	2.6×10^{16}	17	6
19–03–1985	5.4	5.2	5.9×10^{16}	24	7
22–03–1985	5.4	5.7	4.9×10^{16}	17	2
09–05–1986	5.6	5.6	24.7×10^{16}	19	8
19–06–1986	5.3	4.8	2.1×10^{16}	20	5
19–06–1986	5.4	5.1	3.5×10^{16}	23	6

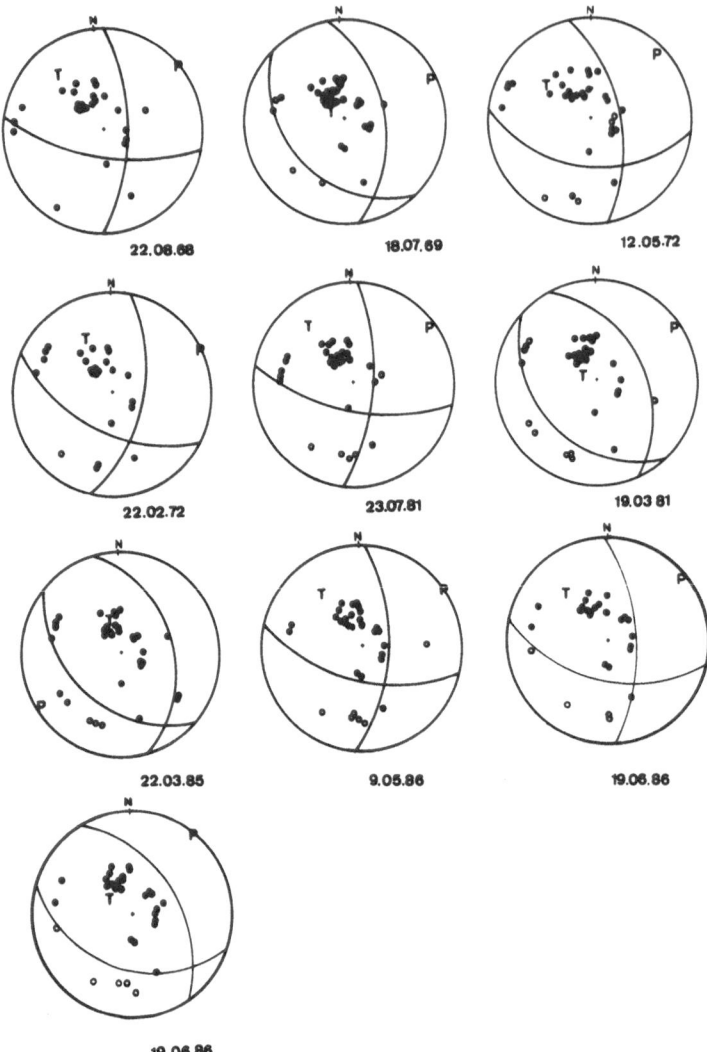

Figure 3

Fault-plane solutions of earthquakes in Bolivia (1968–1986). Parameters are given in Table 2. Diagrams show the lower hemisphere of the focal sphere with black circles as compressions and white circles as dilatations.

The wave-form analysis was used to study the event 2 (LANGSTON and HELMBERGER, 1975; MALGRANGE, and MADARIAGA, 1983). Data correspond to P-waves recorded at long-period WWSSN stations at epicentral distances greater than 30° (Figure 4). The observed P-waves are of very simple form, with a single pulse, and were modeled by source of 6 second duration located at 8 km depth. This result confirms the shallow depth of the foci in this area. The orientation of the

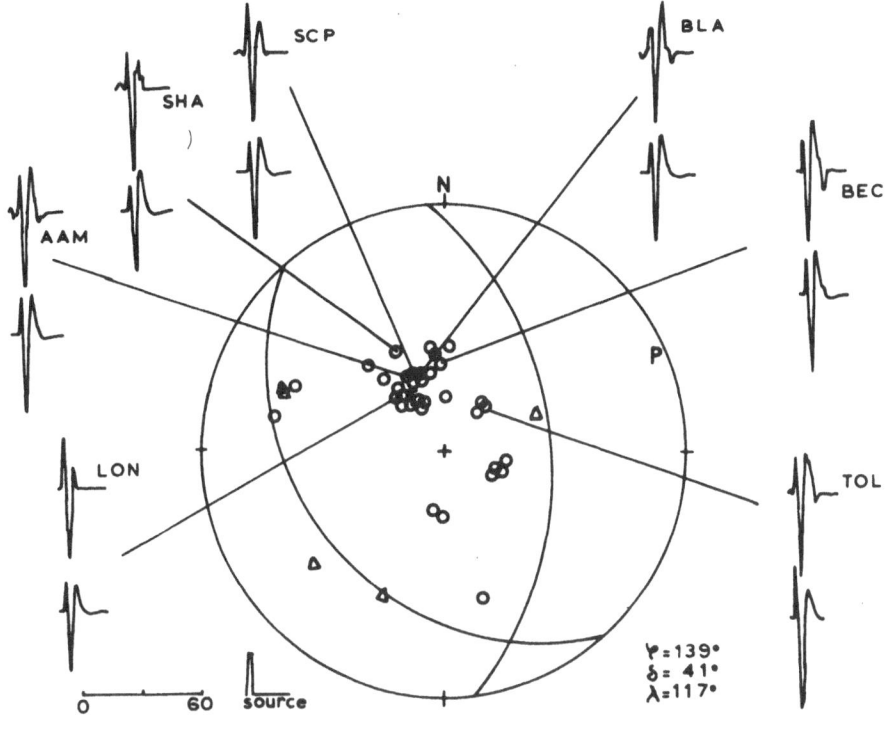

Figure 4

Synthetic and observed *P*-wave-forms of long period WWSSN seismograms for the event 18 July 1969.

mechanism is the same as that derived from the signs of first motions and given in Table 2.

Seismotectonic Analysis

Main geological features of the region and the focal mechanisms for the ten earthquakes studied in this paper are shown in Figure 5. The main geological features are a series of parallel thrust faults which follows the trend of the Sub-Andean Sierra. This trend changes direction at about 17°S, from NW-SE to N-S, at the Santa Cruz bend. Besides these main thrust faults there are other faults, one set with strike E-W and another N-S. Of the 10 mechanisms determined, two are located at the most western thrust faults and the rest at the eastern faults. With respect to the Santa Cruz bend, 6 shocks are located to the north and 4 to the south. Of the 6 shocks to the north, event 10 has a solution corresponding to reverse faulting with purely vertical motion on planes oriented NW-SE, in agree-

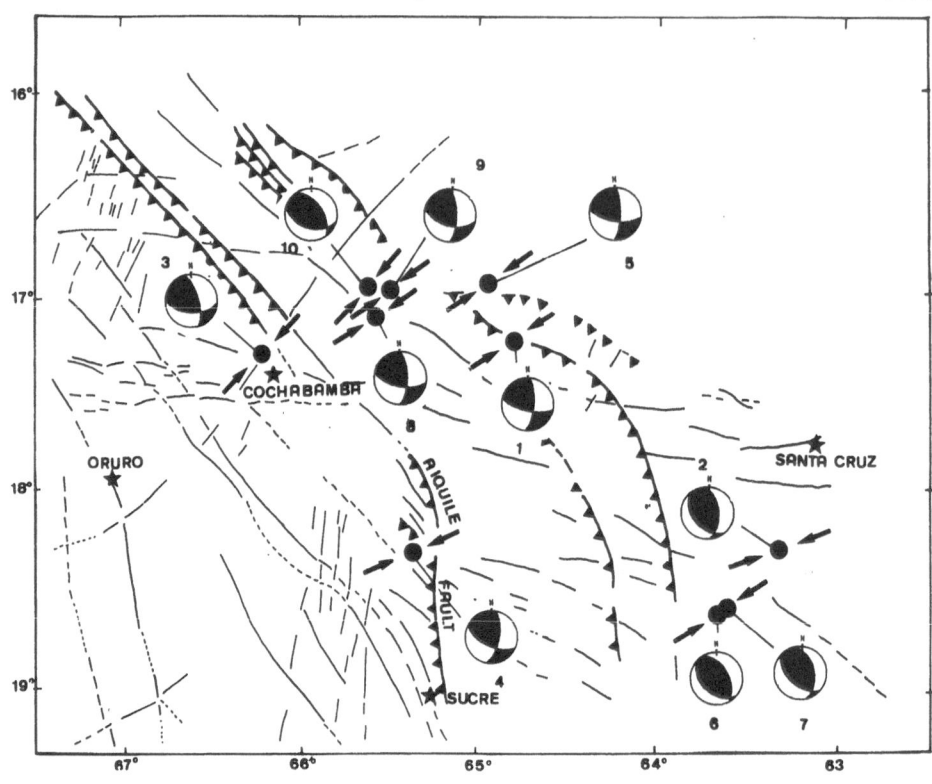

Figure 5
Main fracture systems and focal mechanism of the solutions in this study. Diagrams represent the lower hemisphere of the focal sphere with shaded quadrants for compressions and white for dilatations. Arrows indicate the directions of the horizontal component of the pressure axes.

ment with the orientation of the thrust faults of the region. The other 5 shocks (1, 3, 5, 8 and 9) have a solution of reverse faults, but with large strike-slip components of motion. This component is of left-lateral character on planes oriented WNW-ESE direction and right-lateral on planes oriented N-S. Of the 4 shocks located to the south of the Santa Cruz bend, shocks 2, 6 and 7 have mechanisms similar to that of event 10, with reverse faulting of purely vertical motion on planes striking NW-SE. Shock 4 has a mechanism with a large strike-slip component of motion similar to that of the 5 shocks at the northern part.

There are, then, two types of mechanisms. One is of reverse faulting with pure vertical motion (shocks 2, 6, 7 and 10) and the other with a large strike-slip component (shocks 1, 3, 4, 5, 8 and 9). For the first type, on geological grounds, one must select as the fault plane, the plane dipping to the east with underthrusting of the west block. For the second type, the choice is more difficult. Since there are E-W striking faults in the region, the choice of the plane with similar orientation and left lateral motion with the southern block moving east, seems more plausible.

The strike-slip mechanism located north of the Santa Cruz bend, is related to the eastward displacement of the southern block, which has caused the bend. South of the bend the motion is predominantly dip-slip on N-S striking reverse faults.

The orientation of the pressure axes for all shocks is very consistent, nearly horizontal and trending, as an average, N56°E. This direction is almost perpendicular to the trend of the Sub-Andean Sierra. Under this regional stress, the N-S to NW-SE faults dipping to the east react with reverse vertical motion and with underthrusting of the west block, while the E-W to WNW-ESE faults, produce strike-slip, left-lateral motion, with the southern block moving to the east.

These results may be interpreted, in a general way, in terms of an intraplate situation, subject to regional horizontal compressional stresses, resulting from the convergence of the Nazca and South America plates. This convergence causes a broad deformation zone inside the South America plate between the Bolivian Altiplano and the Brazilian shield that is the region of the Sub-Andean Sierra and East Andean Cordillera. The seismicity and the focal mechanism of earthquakes of the area can, then, be interpreted as produced by an intraplate compression in a crust susceptible to tectonic deformations (JORDAN et al., 1983) which results in a shortening of the crust (SHEFFELS, 1988). In the Santa Cruz bend there is a concentration of tectonic activity responsible for the continuous shallow seismic activity observed in this region. This bend may be considered as caused by an eastward displacement of the block south of Santa Cruz along faults striking E-W to WNW-ESE, as evidenced by the focal mechanisms. South of the bend faulting occurs at N-S striking planes with underthrusting of the west block.

Acknowledgements

The authors wish to thank Dr. R. Cabré, Director of the San Calixto Observatory in La Paz (Bolivia) and Prof. A. Udías, of Universidad Complutense, Madrid (Spain), who read and corrected the manuscript. Part of this work has been supported by the Observatorio de San Calixto, La Paz (Bolivia) and by the Instituto de Cooperación Ibero-Americano of the Spanish government. Publication no. 338 Departamento de Geofísica, Universidad Complutense, Madrid.

REFERENCES

BRILLINGER, D., UDÍAS, A., and BOLT, B. A. (1980), A Probability Model for Regional Focal Mechanism Solution. Bull. Seismol. Soc. Am. 70, 140–170.

BEN-MENAHEN, A., ROSENMAN, M., and HARKRIDER D. G., (1970), Fast Evaluation of Source Parameters from Isolated Surface-wave Signals, Bull. Seismol. Soc. Am. 60, 1337–1387.

CABRÉ, R. (1981), Sismo del 23 de Julio de 1981. Informe para Defensa Civil de Bolivia, Obs. San Calixto, La Paz.

CABRÉ, R., and VEGA, A. (1989), *Sismicidad de Bolivia*, Observatorio de San Calixto, Publication 40, 1–73.

DESCOTES, P. M., and CABRÉ, R. (1965), *Historia sísmica de Bolivia*, Bol. Inst. Boliviano del Petróleo, La Paz, *5*, 16–28.

DESCOTES, P. M., and CABRÉ, R. (1973), *Historia sísmica de Bolivia*, Geofísica Panamericana, IPGH *2*, 251–278.

JAMES, D. E. (1971), *Plate Tectonic Model for Evolution of the Central Andes*, Bull. Geol. Soc. Am. *82*, 3325–3346.

JORDAN, T. E., ISACKS, B. L., ALLMENDIGER, R. W., BREWER, J. A., RAMOS, V. A., and ANDO, C. J., *Andean Tectonics Related to Geometry of Subducted Nazca Plate*, Bull. Geol. Soc. Am. *94*, 1073–1095.

LANGSTON, C. A., and HELMBERGER, D. V. (1975), *A Procedure for Modelling Shallow Dislocation Sources*, Geophys. J. R. Astr. Soc. *42*, 117–130.

MALGRANGE, M., and MADARIAGA, R. (1983), *Complex Distribution of Large Thrust and Normal Fault Earthquakes in the Chilean Subduction Zone*, Geophys. J. R. Astr. Soc. *73*, 489–506.

OCOLA, L. (1984), *Catálogos Sísmicos*, Proyecto SISAN, vol. 1, Lima, Perú.

RODRIGUEZ, R., and VEGA, A. (1976), *El sismo de 22 de febrero de 1976, al sur de Cochabamba*, Publication 27, Obs. San Calixto, La Paz.

SHEFFELS, B.M. (1988), *Structural Constraints on Crustal Shortening in the Bolivian Andean*, Ph. D. Thesis, M.I.T., 170 pp.

VEGA, A. (1976), *El sismo de Arque del 30 de Junio de 1976*, Informe para Defensa Civil de Bolivia, Obs. San Calixto, La Paz.

VEGA, A. (1985), *Catalog of Earthquakes for South America* (SISRA Project), CERESIS (Ed) Vol. 3, 26 pp, Lima, Perú.

VEGA, A. (1986), *El sismo del 9 de Mayo de 1986 en el Chapare, Cochabamba*, Informe para Defensa Civil de Bolivia, Obs. San Calixto, La Paz.

VEGA, A. (1987), *El sismo de Forestal en Santa Cruz*. Informe para Defensa Civil de Bolivia, Obs. San Calixto, La Paz.

VEGA, A. (1989), *Sismicidad Reciente en Bolivia Central*, Meeting of Geophysical Brazilian Society, Rio de Janeiro, Brazil.

(Received February 8, 1991, revised/accepted May 4, 1991)

PAGEOPH, Vol. 136, No. 4 (1991)

0033–4553/91/040459–19$1.50 + 0.20/0
© 1991 Birkhäuser Verlag, Basel

Partial Breaking of a Mature Seismic Gap: The 1987 Earthquakes in New Britain

R. DMOWSKA,[1,2] L. C. LOVISON-GOLOB[1] and J. J. DUREK[2]

Abstract — To better understand the mechanics of subduction and the process of breaking a mature seismic gap, we study seismic activity along the western New Britain subduction segment (147°E–151°E, 4°S–8°S) through earthquakes with $m_b \geq 5.0$ in the outer-rise, the upper area of subducting slab and at intermediate depths to 250 km, from January 1964 to December 1990. The segment last broke fully in large earthquakes of December 28, 1945 ($M_s = 7.9$) and May 6, 1947 ($M_s = 7.7$), and its higher seismic potential has been recognized by McCANN *et al.* (1979). Recently the segment broke partially in two smaller events of February 8, 1987 ($M_s = 7.4$) and October 16, 1987 ($M_s = 7.4$), leaving still unbroken areas.

We observe from focal mechanisms that the outer-rise along the whole segment was under pronounced compression from the late 60's to at least October 1987 (with exception of the tensional earthquake of December 11, 1985), signifying the mature stage of the earthquake cycle. Simultaneously the slab at intermediate depths below 40 km was under tension before the earthquake of October 16, 1987. That event, with a smooth rupture lasting 32 sec, rupture velocity of 2.0 km/sec, extent of approximately 70 km and moment of 1.2×10^{27} dyne-cm, did not change significantly the compressive state of stress in the outer-rise of that segment. The earthquake did not fill the gap completely and this segment is still capable of rupturing either in an earthquake which would fill the gap between the 1987 and 1971 events, or in a larger magnitude event ($M_s = 7.7–7.9$), comparable to earthquakes observed in that segment in 1906, 1945 and 1947.

Key words: Subduction zone, New Britain earthquakes, mature seismic gap, earthquake prediction.

Introduction

Recent observational and theoretical work on earthquake cycles in subduction zones (CHRISTENSEN and RUFF, 1983, 1988; ASTIZ and KANAMORI, 1986; DMOWSKA *et al.*, 1988; DMOWSKA and LOVISON, 1988; ASTIZ *et al.*, 1988; LAY *et al.*, 1989) has explained certain seismic phenomena in relation to stress accumulation and release associated with great underthrust events. It has been realized that temporal variations of stress, associated with earthquake cycles, occur in the subducting slab and, as well, in the area of the outer-rise, oceanward from the main zones of subduction earthquakes. In the outer-rise, the bending stresses present

---

[1] Division of Applied Sciences, Harvard University, Cambridge, MA 02138, U.S.A.
[2] Department of Earth and Planetary Sciences, Harvard University, Cambridge, MA 02138, U.S.A.

become overprinted with tensional stresses in the beginning of the cycle, caused by the slip in the main subduction event. By the latter part of the cycle that has changed to a compressional overprint, occurring because the main thrust zone remains locked while converging motion of the remote ocean floor continues. These factors result in typical tensional outer-rise earthquakes following large subduction events, as well as sporadic compressional ones preceding large subduction events, as documented in the works cited above.

At intermediate depths, in the down-going subducting slab, the tensional stresses caused by slab pull receive a superposed compressional component in the beginning of the cycle, caused by the slip in the main thrust subduction event. In the latter part of the cycle the continuing slab pull and the locking of main thrust zone result in higher tensional stresses at intermediate depths.

We have utilized these new insights when analyzing the mechanics of partial breaking of a mature seismic gap along the western New Britain subduction segment (147°E-152°E, 4°S–8°S). In particular, we have used seismic mechanisms to infer the temporal and spatial changes in stress patterns in the outer-rise and in the downgoing slab caused by the recent partial breaking of the segment in the October 16, 1987 ($m_b = 5.9$, $M_s = 7.4$) earthquake. We have complemented our study by the analysis of the source process of that earthquake.

Tectonic Setting and Seismic History of the New Britain Subduction Segment

The New Britain-Solomon Islands area, east of New Guinea (Figure 1), is characterized by complex interactions between several microplates and high seismicity levels both at the plate interfaces and at intermediate and great depths in the subducted slabs.

The Solomon plate subducts towards NNW under the Bismarck plate along the New Britain trench and towards the northeast under the Pacific plate along the Solomon trench, with a contortion around the junction of the Solomon and New Britain trenches.

In the north, backarc spreading operates along the Bismarck Sea seismic zone, with left-lateral strike-slip movements. This seismic lineation arches from around 2°S, 147°E and enters the Solomon Sea tangentially to the Solomon trench at the convergent region of the Solomon and New Britain trenches, creating in that region a triple junction. Seismic activity in the Bismarck Sea seismic zone is high, and earthquakes with magnitudes larger than 7.0 have occurred there during the past 60 years.

In the south another spreading center is operational along the Woodlark ridge, with weak, shallow seismicity (PASCAL, 1979; LAY and KANAMORI, 1980; COOPER and TAYLOR, 1987).

The Wadati-Benioff zone to the west of the junction between the New Britain and Solomon trenches trends southwest and dips to the northwest at about 70°–75°

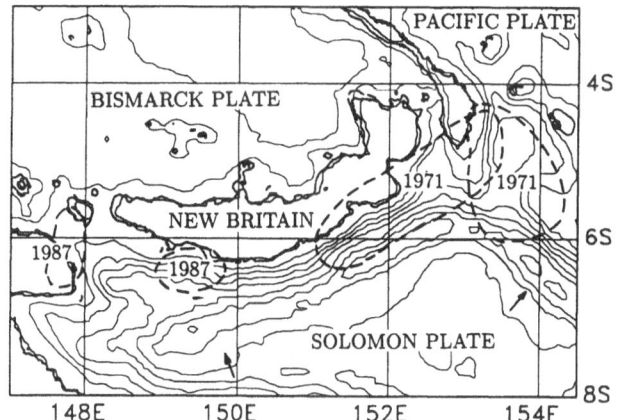

Figure 1
The New Britain subduction segment with aftershock zones of recent large subduction earthquakes (broken line).

(BURBACH and FROHLICH, 1986). Seismicity extends to about 550 km depth, but a significant gap in activity occurs between about 300 and 500 km. At shallow depth the dip of the slab is around 30° (JARRARD, 1986).

To the west of around 148°E the New Britain trench intersects New Guinea and the seismicity forms two planar zones, one dipping south, and the other dipping north (BURBACH and FROHLICH, 1986). The northward dipping zone extends to over a 200 km depth and appears to be a continuous extension of the zone east of 148°E. The zone dipping south extends to nearly a 200 km depth. Both the northward and southward dipping seismic zones terminate to the west at around 145°E.

The New Britain subduction segment is moderately coupled, with variations in rupture extent and occasional ruptures reaching 500 km length, with close clustering of large events and doublets (KANAMORI, 1981; CHRISTENSEN and RUFF, 1988).

In eastern New Britain (151°E-153°E) earlier sequences of events (February 2, 1920, $M_s = 7.9$, September 29, 1946, $M_s = 7.7$, April 23, 1953, $M_s = 7.6$, and July 26, 1971, $M_s = 7.7$) have repeatedly ruptured substantial portions of the plate boundary. The recurrence times for this region appear to be very short, only 25 ± 5 y (LAY and KANAMORI, 1980). Given that the last event occurred in 1971 (a doublet; the aftershock zones are shown in Figure 1, see also SCHWARTZ et al., 1989), the probability of a large earthquake occurring in that area in the next 10 years is estimated at the 59% level (NISHENKO, 1991).

In western New Britain two large earthquakes occurred in 1945 and 1947 (December, 28, 1945, $M_s = 7.9$ and May 6, 1947, $M_s = 7.7$, Table 1, epicenters shown in Figure 2). An event on September 14, 1906 ($M_s = 7.7$, Table 1, Figure 2) is located in this area, but could not be reliably relocated (NISHENKO, 1991). If we

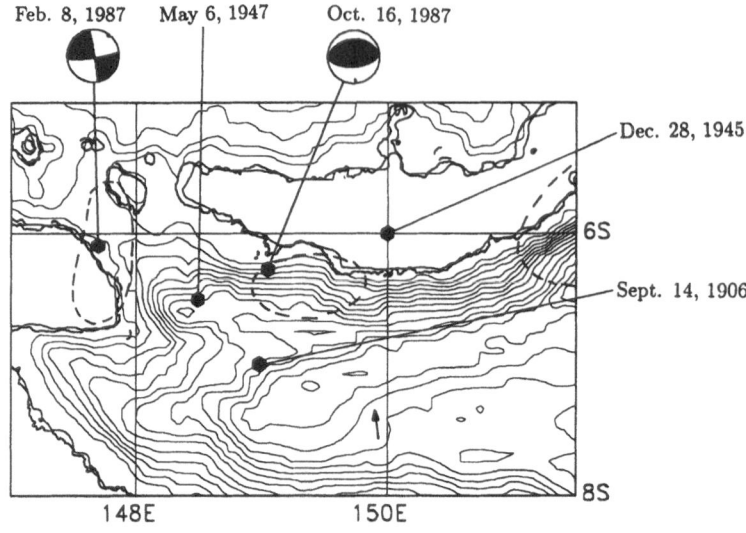

Figure 2

The western New Britain segment with epicenters of large subduction events that occurred in this century. Aftershock zones of recent events shown by broken line. Also shown the CMT solutions for both 1987 earthquakes with $M_s = 7.4$. The direction of plate motion marked by an arrow.

assume that this earlier event was the predecessor to the events in the 1940's, the repeat time is approximately 40 years. The higher seismic potential of western New Britain, based on historic seismicity, has been recognized by McCANN *et al.* (1979) in their map of seismic potential of major plate boundaries. Currently NISHENKO (1991) estimated the probability of the recurrence of a large earthquake in that area within the next 10 years at the 58% level.

Even if the western New Britain segment did not experience any earthquakes with $M_s \geq 7.5$ since the 1945/1947 events, it ruptured partially in 1987 in two

Table 1

Largest shallow earthquakes in western New Britain in this century

Date	Lat. (°)	Lon. (°)	Depth (km)	m_b	M_s	F. M.	Ref.
Sept. 14, 1906	7.00S	149.00E			8.1		GR
Dec. 28, 1945	6.00S	150.00E			7.8		GR
May 6, 1947	6.50S	148.50E	30		7.5		AS87
Feb. 8, 1987	5.94S	147.79E	34	5.8	7.4	S	ISC/HAR
Oct. 16, 1987	6.25S	149.09E	24	5.9	7.4	C	PDE/HAR

AS87—ASTIZ, 1987; GR—magnitude from GUTENBERG and RICHTER, 1949; HAR—centroid moment tensor solutions by the Harvard group; ISC—Bullet. Intern. Seismol. Center; PDE—Prelim. Deter. of Epic., USGS; C—compr., S—strike-slip focal mechanisms.

earthquakes with $M_s = 7.4$ (February 8, 1987, $m_b = 5.8$ and October 16, 1987, $m_b = 5.9$, Table 1; the 10-day aftershock zones are shown in Figure 1 and focal mechanisms in Figure 2). These earthquakes certainly did not fill the gap recognized by McCANN et al. (1979) and NISHENKO (1991). Their size, however, was significant enough for us to decide to look into the subduction of that part of the New Britain segment, with particular attention given to the outer-rise and interme-diate depth earthquakes and their mechanisms, with the aim to better understand the mechanics of partial breaking of a mature seismic gap. A preliminary report of our work has been presented earlier (LOVISON et al., 1988).

The 1987 ($M_s = 7.4$) Earthquakes in New Britain

Two earthquakes with $M_s = 7.4$ occurred in western New Britain in 1987.

The first one ruptured the western end of the New Britain subduction segment on February 8, 1987 ($m_b = 5.8$, hypocentral depth 34 km, 10-day aftershock zone and mechanism shown in Figure 2). It was a strike-slip event with a slight reverse component. The length of the ruptured area was approximately 120 km, and the width 45 km.

The other $M_s = 7.4$ earthquake occurred on October 16, 1987 ($m_b = 5.9$, 10-day aftershock zone and mechanism shown in Figure 2) in the middle of the western segment of New Britain, recognized as a seismic gap by McCANN et al. (1979) and NISHENKO (1991). It was a shallow, thrust event; the length of the rupture zone was approximately 100 km, and the width 60 km, the duration of rupture process was 32 seconds. The distribution of the few catalogued aftershocks suggested a dominant component of rupture to the east-northeast, but poorly constrained the extent of rupture. We have analyzed the rupture process of that earthquake and the results are presented in the next section.

Analysis of the Source Process of the October 16, 1987 New Britain Earthquake

We have applied recently developed source inversion methods, which resolve spatial and temporal variations in the rupture process. The particular inversion method employed, the very broad-band (VBB) technique developed by EKSTRÖM (1987, 1989), combines the analysis of broad-band P-wave phases with results of the centroid-moment tensor (CMT) inversion of long-period body and mantle wave seismograms (DZIEWONSKI et al., 1981). Because an earthquake source is com-pletely described by the moment-density rate tensor $\dot{m}_{ij}(\mathbf{x}, t)$, its determination is the objective of the inversion process. However, the actual spatial and temporal distribution of the moment rate tensor is difficult to constrain with a finite number of observations, forcing an approximation of $\dot{m}_{ij}(\mathbf{x}, t)$ to be adopted. A parameterization of the moment density rate tensor, which is enlightening while remaining practical to apply, consists of a linearly propagating point source. The seismic source is represented by a focal mechanism, given by the normalized zero-order moment tensor, a source time function describing the rate of moment release, and a point source initially at location x_0, with a directivity vector v:

$$\dot{m}_{ij}(\mathbf{x}, t) = \{M_{ij}/M_0\} \times F(t - t_0) \times \delta(\mathbf{x} - \mathbf{x}_0 - (t - t_0)\mathbf{v}). \tag{1}$$

Such a parameterization approximately characterizes all but the most complex of sources, in which case additional model complexity must be introduced (see EKSTRÖM, 1987).

In contrast to other studies, the VBB technique combines GDSN long-period and triggered short-period records and deconvolves for broad-band ground displacement (flat response from 1 to 100 Hz) to form the set of observations to be inverted. The obvious advantage of using displacement records is that $\dot{m}_{ij}(\mathbf{x}, t)$ is directly related to ground displacement. The arrival times are manually picked to provide proper alignment of the observed and synthetic seismograms, a necessary process which unfortunately removes travel time information about the epicentral location. The focal depth, however, is constrained by the shapes and arrival times of the depth phases pP and sP given an accurate model for the structure of the source region. During the inversion, the source parameters defined in equation (1), as well as the focal depth, are determined using a standard iterative least-squares technique in which the robust CMT long-period mechanism is employed as a weighted constraint.

The CMT solution for the October 16, 1987 event indicates a shallow thrust mechanism which is consistent with the tectonic regime. The predicted CMT depth of 48 km seems too deep for the region. It has been observed (EKSTRÖM, 1987) that the CMT inversion tends to overestimate the focal depth in certain geographical regions, most likely the result of high-frequency deviations of the source region structure from the weakly heterogeneous earth model used for calculation of the synthetic seismograms in the CMT analysis.

In applying the VBB method, ten deconvolved records were available. The model for the structure of the source region includes a two-layer crust with a 3 km surface water layer (Table 2). The final source solution is presented in Figure 3 and possesses a normalized residual variance of only 0.046. Records from the two stations SNZO and NWAO, which lie near the nodal plane, were not included in the final inversion. The predicted seismograms were calculated for these stations, however, and the general agreement in shape is good, considering the reduced signal-to-noise ratio near the nodal plane. The remaining stations reveal simple pulses indicative of an unusually uniform rupture. The inversion corroborates the smooth rupture process, retrieving a source time function with a dominant low frequency component. The temporal extent of the rupture was 32 seconds, although the source time function was unconstrained up to 40 seconds. The smooth source time

Table 2

Crustal velocity structure used in the VBB calculation of synthetic seismograms

Depth range (km)	P-wave velocity (km/s)	S-wave velocity (km/s)	Density (g/cm³)
0.0–3.0	1.52	0.00	1.02
3.0–15.0	5.80	3.20	2.60
15.0–24.4	6.80	3.90	2.90
24.4–64.4	8.10	4.50	3.40
64.4		PREM	

function suggests that the fault rupture was nearly uniform with no significant asperities or barriers to modify the rupture propagation. The fault plane solution does not significantly deviate from the CMT solution of a shallow thrust mechanism (Table 3). The moment of 1.2×10^{27} dyne-cm determined by the VBB inversion is consistent with that determined by the CMT inversion, while the resolved focal depth of 36 km is appreciably shallower than that of the CMT solution and more consistent with the regional tectonics. However, when investigating a large event the wave forms are more dominated by the shape of the source time function and the depth phases inverted to constrain the depth are suppressed by direct arrivals from the source. In such a case the VBB inversion experiences a diminished sensitivity to depth. Further inversions with a depth fixed at 25 km resulted in marginally poorer fits and unacceptable alterations of the selected arrival times, allowing the conclusion that the focal depth was not likely to be less than 30 km.

The directivity vector describing the centroid propagation has a trend of N59°E, a plunge of 26°, and is essentially contained within the fault plane. The rupture velocity of 2.0 km/sec, combined with the source duration of 32 seconds, requires a lower limit of approximately 60 km of breakage. Both the direction and extent of rupture obtained from the VBB analysis are consistent with the limited aftershock distribution.

Table 3

CMT and VBB moment tensor comparisons

	Moment Tensor ($\times 10^{27}$ dyne–cm)	
	CMT solution	VBB solution
$M_{\tau\tau}$	1.15	1.13
$M_{\theta\theta}$	−1.11	−1.10
$M_{\phi\phi}$	−0.05	−0.04
$M_{\tau\theta}$	0.54	0.50
$M_{\tau\phi}$	0.04	−0.04
$M_{\phi\phi}$	−0.07	−0.08

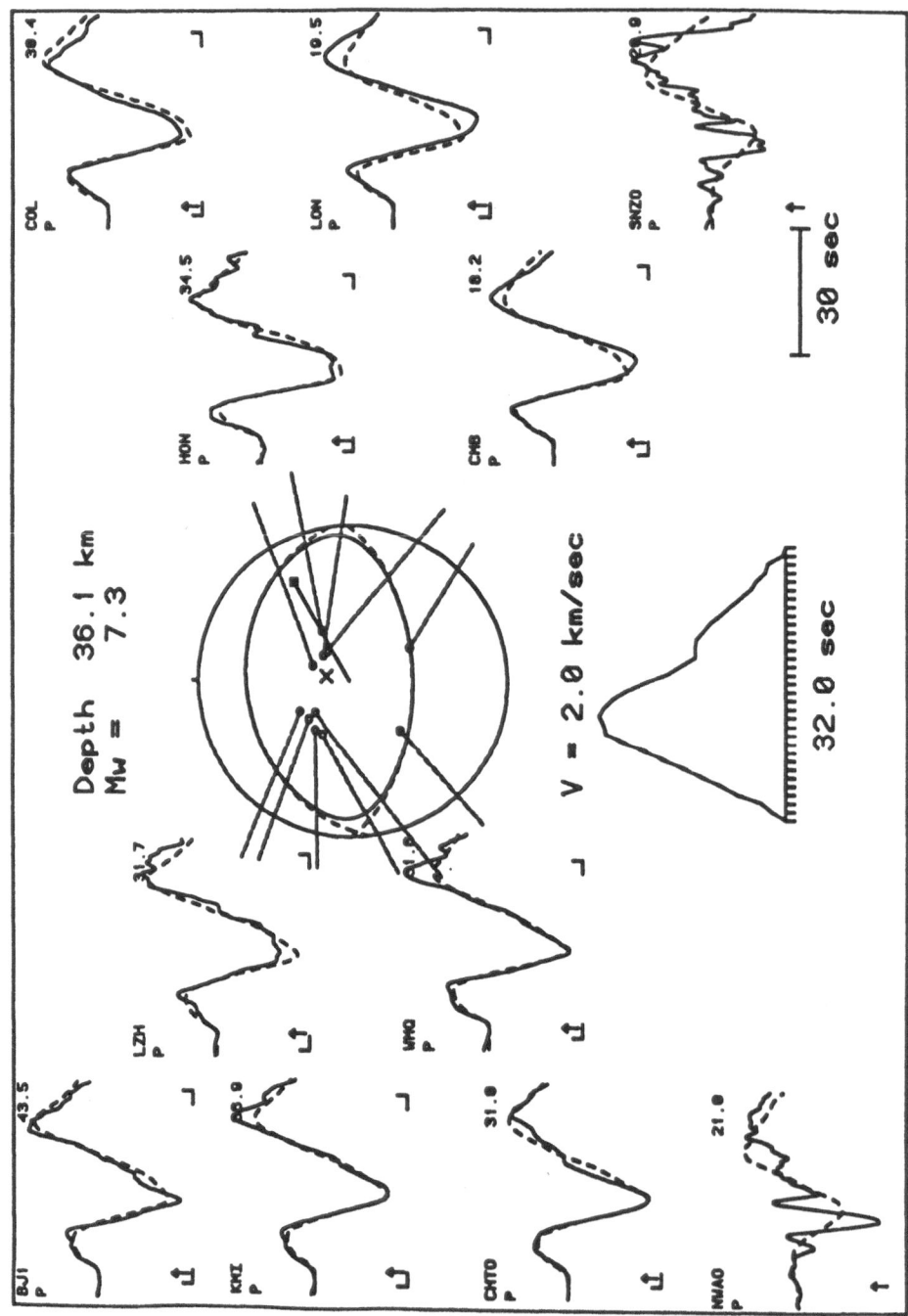

Mechanical Behavior of the Outer-Rise

In order to analyze any temporal and/or spatial changes in the state of stress in the outer-rise area of the western New Britain, we show the seismicity in that area in Figures 4 and 5. The CMT solutions are presented when available. All earthquakes are listed in Tables 4 and 6, with mechanisms, if known.

Figure 4 presents the earthquakes with $m_b \geq 5.0$ for the period between January 1, 1964 and October, 16, 1987, that is for as long time as possible before the October 1987 earthquake which partially ruptured that segment. The data for the period January 1, 1964 to July 31, 1987 are taken from the ISC catalogue, and for August 1, 1987 to October 16, 1987 from the USGS PDE catalogue. We are aware of the differences in confidence levels of these catalogues. However, we would like to cover the longest time periods possible.

The compressional state of stress, characteristic of the outer-rise areas adjacent to mature seismic gaps (DMOWSKA et al., 1988; DMOWSKA and LOVISON, 1988; CHRISTENSEN and RUFF, 1988; ASTIZ et al., 1988; LAY et al., 1989) has been

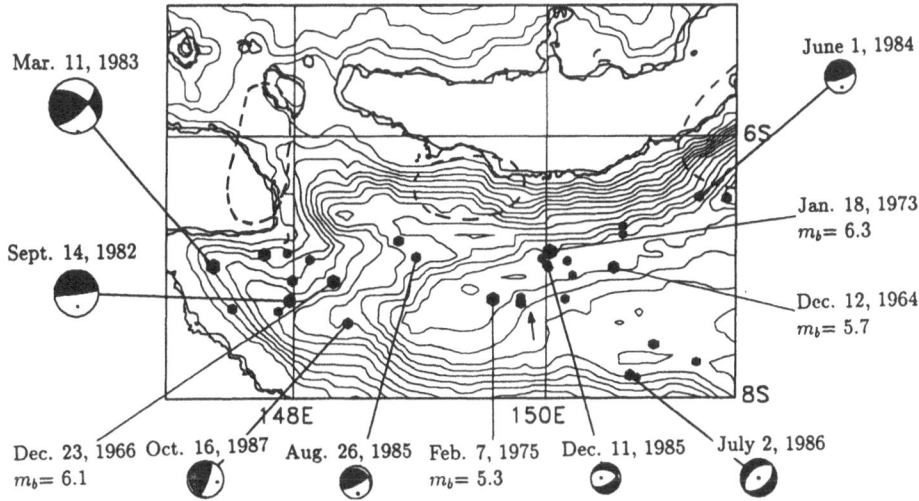

Figure 4

Epicenters of earthquakes with $m_b \geq 5.0$ in the outer-rise area of western New Britain for period January 1, 1964 to October 16, 1987.

Figure 3

Source solution for the New Britain event of October 16, 1987 as obtained by the VBB analysis of EKSTRÖM (1987). The source mechanism is shown in lower hemisphere projection. For each station, the observed record is solid, while the synthetic seismogram is dotted. The arrows indicate the selected P-wave arrival times and brackets the data windows used for inversion. The directivity vector is plotted on the focal sphere with the arrow denoting the direction of centroid propagation.

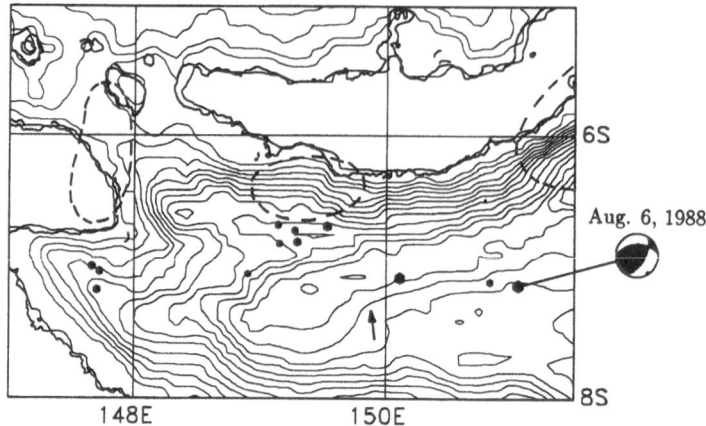

Figure 5
Epicenters of earthquakes with $m_b \geq 4.5$ in the outer-rise area of western New Britain for period October 16, 1987 to December 31, 1990.

recognized in that segment by CHRISTENSEN and RUFF (1988), based on two compressional earthquakes, one in the western end of the trench area, on December 23, 1966 ($m_b = 6.1$), and the other in the middle of the segment, on January 18, 1973 ($m_b = 6.3$). We can add to that list two other compressional events that occurred later, namely the June 1, 1984 event ($m_b = 5.2$), and the August 26, 1985 event ($m_b = 5.2$). Both earthquakes are shown in Figure 4 with their CMT solutions. All these events suggest that before the October 16, 1987 earthquake, that partially ruptured the area, the outer-rise has been indeed in the compression dominated state of stress along the whole western segment of the New Britain, signaling the maturity of that segment.

Because of the tectonic complexity of the western end of the New Britain trench, that is the area around 147°E–148°E, we do not attempt to interpret the mechanisms of earthquakes in that area (Figure 4).

We comment here on an outer-rise earthquake that slightly puzzled CHRISTENSEN and RUFF (1988) during their analysis of the state of stress in the New Britain segment. This is the tensional event of December 11, 1985 (shown with its CMT solution in Figure 4), that occurred near the January 18, 1973 compressional earthquake. CHRISTENSEN and RUFF (1988) suggested that the accumulated compressional stress has been dissipated by some mechanism such as aseismic slip in the subduction zone following the compressional outer-rise event. We have found that a cluster of other tensional earthquakes with $5.0 \leq m_b \leq 5.7$ occurred during a few days in September 1985, that is a few months earlier than the outer-rise event, in or close to the thrust contact zone adjacent to the outer-rise earthquake. The cluster is shown in Figure 6, with CMT solutions when known, and the earthquakes are listed in Table 5.

Table 4

Earthquakes with $m_b \geq 5.0$ in the outer-rise area of western New Britain, January 1, 1964–October 16, 1987

Date	Lat. (°)	Lon. (°)	Depth (km)	m_b	M_s	F. M.	Ref.
Dec. 12, 1964	7.00S	150.54E	33	5.7			ISC
Aug. 26, 1966	7.31S	147.52E	60	5.3			ISC
Dec. 23, 1966	7.11S	148.31E	46	6.1		C	ISC/JM
Dec. 30, 1967	6.69S	150.61E	26	5.0			ISC
Feb. 26, 1970	6.95S	150.17E	37	5.0			ISC
Apr. 30, 1971	7.24S	150.15E	27	5.1			ISC
Jan. 18, 1973	6.88S	150.03E	38	6.3		C	ISC/P79
Feb. 23, 1973	6.98S	150.01E	40	5.2			ISC
Dec. 4, 1974	7.33S	147.88E	60	5.0			ISC
Feb. 7, 1975	7.24S	149.58E	9	6.2			ISC
Feb. 7, 1975	7.23S	149.80E	15	5.3			ISC
Feb. 25, 1975	7.27S	149.80E	21	5.5			ISC
Aug. 14, 1975	6.90S	147.77E	53	5.8			ISC
June 9, 1976	6.47S	151.43E	21	5.5			ISC
Aug. 19, 1976	7.06S	150.21E	36	5.0			ISC
May 10, 1977	7.73S	151.19E	13	5.1			ISC
Oct. 14, 1980	6.80S	148.83E	59	5.3			ISC
June 10, 1981	6.94S	148.13E	50	5.2	5.0		ISC
Sept. 14, 1982	7.25S	147.97E	34	5.3	6.3	C	ISC/HAR
Nov. 28, 1982	6.75S	150.61E	33	5.0			ISC
Mar. 11, 1983	6.99S	147.37E	66	5.9		C	ISC/HAR
June 1, 1984	6.46S	151.21E	16	5.2		C	HAR
June 6, 1984	6.89S	147.95E	53	5.0			ISC
June 9, 1984	7.10S	147.99E	57	5.4			ISC
Jan. 9, 1985	6.50S	150.50E	11	5.3	4.5		ISC
Mar. 16, 1985	6.93S	149.97E	51	5.0			ISC
Aug. 26, 1985	6.92S	148.97E	33	5.2	6.1	C-S	ISC/HAR
Dec. 11, 1985	7.00S	150.02E	29	5.3		T	ISC/HAR
May 27, 1986	7.85S	150.73E	33	5.7	4.6		ISC
May. 31, 1986	7.59S	150.86E	10	5.4			ISC
July. 2, 1986	7.83S	150.67E	19	5.5	4.8	T	ISC/HAR
Oct. 16, 1987	7.42S	148.43E	39	5.4	4.5	C-S	PDE/HAR

HAR—centroid moment tensor solutions by the Harvard group; ISC—Bullet. Intern. Seismol. Center; JM—JOHNSON and MOLNAR, 1972; P79—PASCAL, 1979; PDE—Prel. Determ. of Epic.— USGS; C—compr., T—tens., S—strike-slip focal mechanisms.

Such tensional earthquakes are unusual in or close to the thrust contact zone, as unusual as a tensional outer-rise earthquake in the area adjacent to a mature seismic gap. The cluster and the outer-rise event are in the same subducting strip of the oceanic crust (see Figures 4 and 6). We infer that it is mechanically plausible that an aseismic slip downdip from that area, perhaps along the lower part of the interface between the slab and the Bismarck plate, caused a temporary tensional state of stress in the area of cluster and in the outer-rise as well. We have inferred

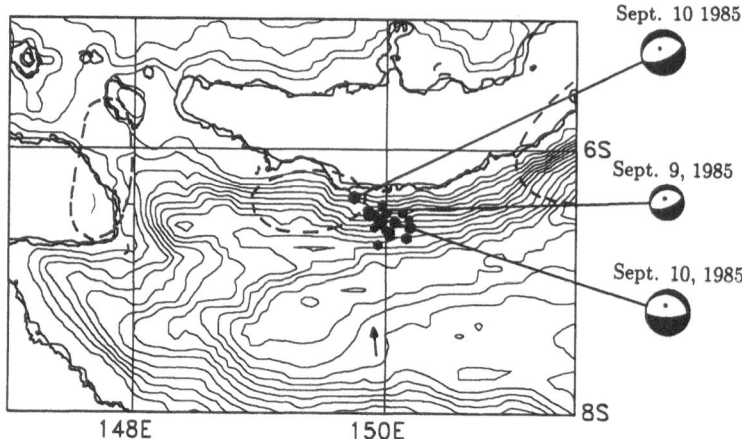

Figure 6
Cluster of tensional earthquakes in or close to the thrust contact zone or in the pretrench area from
September 1985.

Table 5

Cluster of earthquakes with $m_b \geq 5.0$ from September 1985

Date	Lat. (°)	Lǿn. (°)	Depth (km)	m_b	M_s	F. M.	Ref.
Sept. 9, 1985	6.47S	149.86E	19	5.5	5.3	T	ISC/HAR
Sept. 10, 1985	6.50S	149.87E	37	5.1	4.8	T	ISC/HAR
Sept. 10, 1985	6.37S	149.75E	10	5.7	6.3	T	ISC/HAR
Sept. 10, 1985	6.53S	149.97E	19	5.0			ISC
Sept. 10, 1985	6.73S	149.94E	49	5.1			ISC
Sept. 10, 1985	6.43S	149.97E	46	5.4	5.1		ISC
Sept. 10, 1985	6.58S	149.97E	33	5.1			ISC
Sept. 10, 1985	6.54S	150.18E	41	5.0			ISC
Sept. 10, 1985	6.56S	150.07E	10	5.4			ISC
Sept. 10, 1985	6.64S	150.03E	10	5.4			ISC
Sept. 10, 1985	6.60S	150.19E	33	5.4			ISC
Sept. 10, 1985	6.68S	150.16E	10	5.3			ISC
Sept. 10, 1985	6.66S	150.03E	10	5.3			ISC
Sept. 11, 1985	6.49S	150.13E	33	5.0			ISC
Sept. 11, 1985	6.62S	150.00E	22	5.1			ISC
Sept. 11, 1985	6.60S	149.91E	33	5.0			ISC
Sept. 12, 1985	6.52S	149.93E	33	5.2			ISC
Sept. 14, 1985	6.49S	150.00E	50	5.1			ISC

HAR—centroid moment tensor solutions by the Harvard group; ISC—Bullet. Intern. Seismol.
Center; T—tens. focal mechanisms.

similar events along the Mexico convergent plate boundary (though they were smaller, causing only temporary quietening of thrust events), and we modelled such interactions as well (DMOWSKA et al., 1988).

The outer-rise earthquakes in the western segment of New Britain after its partial breaking, that is after October 16, 1987, are shown in Figure 5 with CMT solutions when available, and are listed in Table 6. The data are taken from the USGS PDE catalogue and cover the period until December 31, 1990, for events with $m_b \geq 4.5$.

There is a cluster of small events in the pretrench area adjacent to the aftershock zone of the October 16, 1987 earthquake, the largest with $m_b = 5.1$. We do not have their mechanisms, but we would expect them to be classical tensional earthquakes that follow a subduction event; many of such earthquakes have been observed in the New Britain and Solomon trenches after the 1971 and 1975 interplate doublets (LAY et al., 1989).

The largest event, shown in Figure 5 with its CMT solution, is a compressional earthquake, with a strike-slip component, in front of the unbroken part of the subduction segment.

We infer from Figure 5, that the state of stress in the outer-rise along the western segment of New Britain did not change dramatically as a result of the partial breaking of the segment and that it is still compressional, typical for the outer-rise areas adjacent to mature seismic gaps (see e.g., DMOWSKA et al., 1988; CHRISTENSEN and RUFF, 1988; DMOWSKA and LOVISON, 1988).

Table 6

Earthquakes with $m_b \geq 4.5$ in the outer-rise area of western New Britain, October 16, 1987–December 31, 1990

Date	Lat. (°)	Lon. (°)	Depth (km)	m_b	M_s	F. M.	Ref.
Oct. 17, 1987	7.05S	148.91E	33	4.5			PDE
Oct. 20, 1987	6.82S	149.16E	33	4.6			PDE
Nov. 2, 1987	6.81S	149.30E	43	4.8			PDE
Jan. 5, 1988	7.17S	147.70E	63	4.9			PDE
Feb. 11, 1988	6.69S	149.54E	23	5.1			PDE
July 5, 1988	6.72S	149.28E	33	4.6			PDE
Aug. 6, 1988	7.14S	151.06E	25	5.9	5.7	C-S	PDE/HAR
Dec. 6, 1988	6.99S	147.66E	56	4.9			PDE
May 28, 1989	7.03S	147.72E	33	4.7			PDE
July 27, 1989	7.11S	150.83E	33	4.6			PDE
Apr. 25, 1990	7.08S	150.11E	2	5.4			PDE
June 16, 1990	6.68S	149.15E	5	4.6			PDE

HAR—centroid moment tensor solutions by the Harvard group; PDE—Prel. Determ. of Epic.—USGS; C—compr., S—strike-slip focal mechanisms.

Mechanical Behavior of the Slab at Intermediate Depths

The New Britain subduction segment is characterized by abundant seismicity at intermediate depths, with numerous tear faults and oblique mechanisms suggesting the complexity of the intraplate stress regime. The steep dip of the subducting slab presents some difficulty in distinguishing intraplate tensional events from interplate thrusts. Also, there is a general tendency for the catalogue source depths of large interplate events in this region to be greater than in other zones. With all these limitations in mind we will analyze here the mechanical behavior of the slab at intermediate depths before and after the partial rupturing of the segment in the October 16, 1987 earthquake.

Table 7

Earthquakes with $m_b \geq 5.7$ at intermediate depths (40 km to 250 km) in western New Britain, January 1, 1964–October 16, 1987

Date	Lat. (°)	Lon. (°)	Depth (km)	m_b	M_s	F. M.	Ref.
Jan. 14, 1964	5.21S	150.83E	169	5.7			ISC
July 31, 1964	6.01S	149.36E	58	5.7			ISC
Nov. 17, 1964	5.75S	150.74E	60	6.3		T	ISC/JM
Dec. 7, 1964	5.39S	151.24E	70	5.7		T	ISC/JM
March 4, 1965	5.46S	147.00E	191	5.7			ISC
Sept. 22, 1965	5.35S	151.49E	62	5.7		S	ISC/C73
Apr. 1, 1966	5.81S	149.19E	97	5.7			ISC
Sept. 16, 1968	6.08S	148.77E	49	5.9			ISC
Mar. 10, 1969	5.60S	147.29E	194	5.7			ISC
Nov. 16, 1970	6.05S	148.55E	80	5.7			ISC
Feb. 27, 1975	6.07S	148.21E	72	5.8			ISC
May 22, 1976	5.60S	148.35E	164	5.9			ISC
Jan. 24, 1978	5.97S	148.87E	95	5.7	5.4	S	ISC/HAR
Oct. 15, 1978	5.61S	148.08E	168	5.7	4.9	C	ISC/HAR
Nov. 1, 1979	5.91S	150.19E	56	5.7	5.5		ISC
Feb. 27, 1980	6.02S	150.12E	52	5.9	6.7	T	ISC/HAR
Feb. 27, 1980	6.04S	150.01E	55	5.8			ISC
Feb. 24, 1981	6.04S	148.77E	79	5.8	6.5	T	ISC/HAR
Aug. 26, 1981	5.34S	151.48E	67	5.9		T	HAR
June 9, 1982	5.66S	150.99E	80	5.8		T	ISC/HAR
Jan. 16, 1983	5.45S	147.05E	228	5.8		S-T	ISC/HAR
Jan. 26, 1983	6.13S	150.05E	47	5.7		S	HAR
May 10, 1983	5.40S	150.94E	116	6.0		S	ISC/HAR
Sept. 30, 1984	6.05S	148.49E	77	5.7		T	ISC/HAR
Dec. 30, 1985	5.53S	150.69E	116	5.8		T	ISC/HAR
Aug. 20, 1986	5.32S	151.33E	92	5.7		S	ISC/HAR
May. 12, 1987	5.30S	151.33E	107	5.7		S-T	ISC/HAR

C73—Curtis, 1973; HAR—centroid moment tensor solutions by the Harvard group; ISC—Bullet. Inter. Seism. Center; JM—Johnson and Molnar, 1972; C—compr., T—tens., S—strike-slip focal mechanisms.

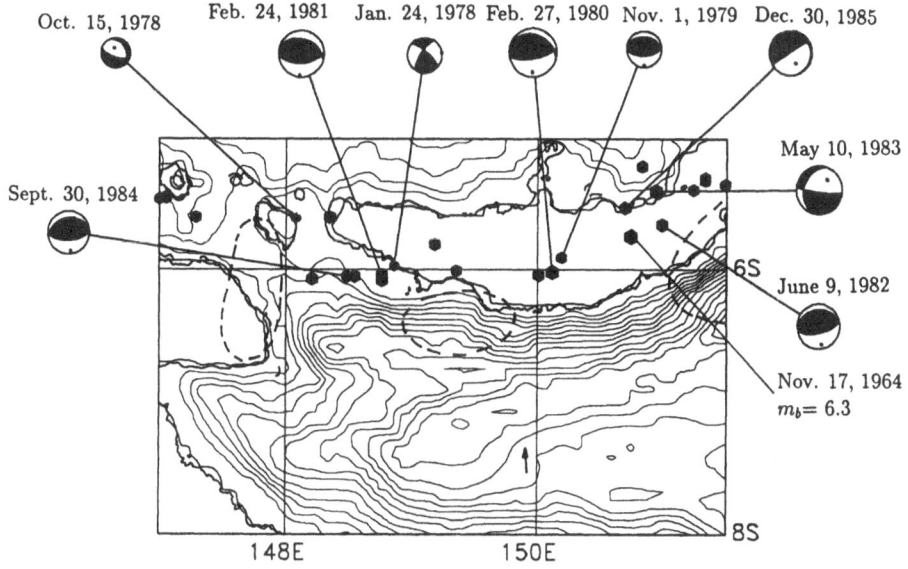

Figure 7

Epicenters of earthquakes with $m_b \geq 5.7$ at intermediate depths (40–250 km) in the area downdip from western New Britain, for the period January 1, 1964 to October 16, 1987.

Because of the abundant seismicity, we are limiting our search to the earth-quakes with $m_b \geq 5.7$ only. Epicenters of such events, located at depths 40 km to 250 km downdip from the western segment of New Britain, during the period between January 1, 1964 and October 16, 1987, are shown in Figure 7 and are listed in Table 7. Data are taken from the ISC catalogues for the period between January 1, 1964 and July 31, 1987, and from the USGS PDE catalogue for the period between August 1, 1987 and October 16, 1987. The CMT solutions of earthquakes located downdip from the segment between the February 8, 1987 aftershock zone and the western end of the 1971 aftershock zone are shown in Figure 7, if available.

As mentioned before, the western segment of New Britain ruptured fully in 1945 and 1947 earthquakes, so, with approximately 40 years recurrence time, we are investigating the mechanical behavior of the slab at intermediate depths for approximately the second half of the current, not yet closed, cycle.

In general the slab at intermediate depths is characterized by the tensional state of stress, as inferred from the mechanisms of larger ($m_b \geq 5.7$) earthquakes (Figure 7). At that level of magnitude there is no observable increase in the number of tensional earthquakes downdip from the October 16, 1987 event, that partially ruptured the segment. However in the 1980's there were many more tensional events at intermediate depths (see Table 7) downdip from the unbroken

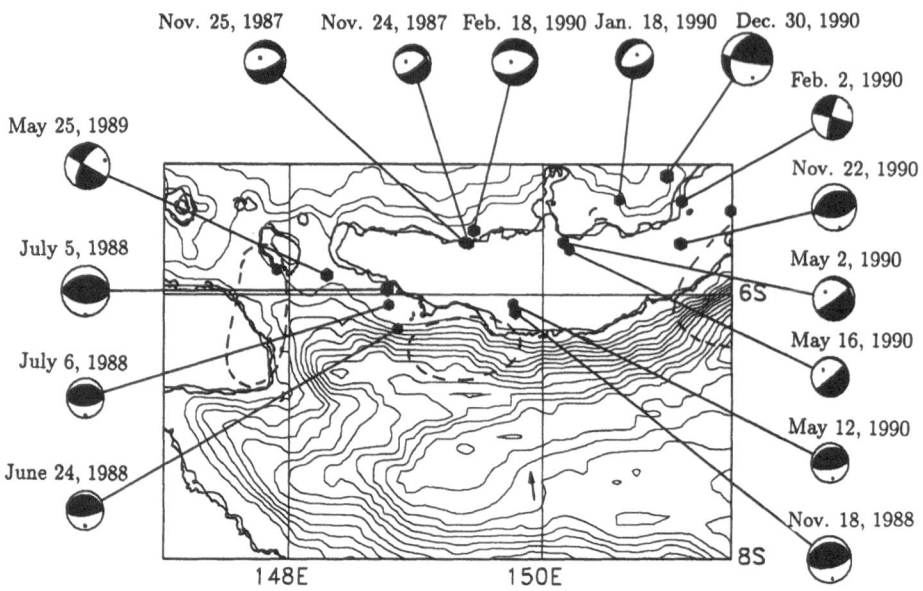

Figure 8

Epicenters of earthquakes with $m_b \geq 5.5$ at intermediate depths (40–250 km) in the area downdip from western New Britain, for the period October 16, 1987 to December 31, 1990.

Table 8

Earthquakes with $m_b \geq 5.5$ at intermediate depths (40 km to 250 km) in western New Britain, October 16, 1987–December 31, 1990

Date	Lat. (°)	Lon. (°)	Depth (km)	m_b	M_s	F. M.	Ref.
Nov. 24, 1987	5.61S	149.42E	142	5.5		C	PDE/HAR
Nov. 25, 1987	5.60S	149.39E	141	5.7		C	PDE/HAR
June 24, 1988	6.26S	148.86E	42	5.5	5.3	C	PDE/HAR
July 5, 1988	5.96S	148.78E	53	6.0	6.8	C	PDE/HAR
July 6, 1988	6.08S	148.79E	67	5.5		C	PDE/HAR
Aug. 5, 1988	5.81S	147.90E	141	5.5			PDE
Nov. 18, 1988	6.13S	149.79E	61	5.8	6.4	C	PDE/HAR
May 25, 1989	5.85S	148.30E	112	5.9		S	PDE/HAR
Jan. 18, 1990	5.28S	150.60E	136	5.6		C	PDE/HAR
Feb. 2, 1990	5.29S	151.10E	44	5.7	5.5	S	PDE/HAR
Feb. 18, 1990	5.51S	149.46E	144	5.9		C	PDE/HAR
May 2, 1990	5.60S	150.16E	82	6.2		C	PDE/HAR
May 12, 1990	6.07E	149.77E	74	5.6		C	PDE/HAR
May 16, 1990	5.66S	150.21E	77	5.6		C	PDE/HAR
Nov. 22, 1990	5.61S	151.09E	45	5.9	6.0	C	PDE*/HAR
Dec. 24, 1990	5.36S	151.49E	49	5.7	5.6	C	PDE*/HAR
Dec. 30, 1990	5.09S	150.98E	188	6.7	7.0	S	PDE*/HAR

HAR—centroid moment tensor solutions by the Harvard group; PDE—Prel. Determ. of Epic.; * weekly list—USGS; C—compr., S—strike-slip.

segment between the October 16, 1987 and 1971 earthquakes (it should be noted here that we interpret the mechanisms according to the direction along the dipping slab). This observation suggests that that part of the subducting slab is under increased tension, typical for areas downdip from a mature seismic gap (ASTIZ et al., 1988; LAY et al., 1989).

Figure 8 shows seismicity at the same depth for the period between October 16, 1987 and December 31, 1990, data being taken from the USGS PDE catalogue. All earthquakes are listed in Table 8 with their mechanisms, if known. Because this is a short period of time, we show seismicity at $m_b \geq 5.5$ level, slightly lower than in Figure 7. In spite of this we could not add any significant new observation, and we must conclude that the slab at intermediate depths downdip from the western segment of New Britain is still under tension.

Discussion and Conclusions

We have investigated in detail the seismicity in the outer-rise and downdip areas adjacent to the western segment of New Britain, which broke fully in the 1945 and 1947 earthquakes, and partially in the October 16, 1987 event. The outer-rise is currently in compression, and the slab at intermediate depths under tension, characteristic of outer-rise and downdip areas adjacent to mature seismic gaps. These observations, added to the fact that the segment has an approximate recurrence time of 40 years (NISHENKO, 1991), suggest that it might rupture in the nearest 10 years or so. Unfortunately our observations concerning the state of stress in the outer-rise and at intermediate depths are of only intermediate-term precursors (see also DMOWSKA and LOVISON, 1988), and we are not able to make this prediction more specific. Also, our observations do not allow us to comment on the future mode of rupture of that segment: it might break in an earthquake filling only the gap between the 1971 and 1987 earthquakes (the length of that segment is approximately 150 km), or the whole western segment could rupture in one event with $M_s = 7.7-7.9$, similar to the magnitudes of the 1945 and 1947 events.

It is plausible that the situation in western New Britain is mechanically similar to that of the Michoacan, Mexico, where the earthquake of Playa Azul on October 25, 1981 ($M_s = 7.3$) occurred in the central part of the area identified previously as a seismic gap (Michoacan gap, KELLEHER et al., 1973) along the Cocos-North American convergent plate boundary, rupturing an area of about 40×20 km^2. Subsequently, the whole gap broke on September 19, 1985 in the $M_s = 8.1$ Michoacan earthquake. It is perhaps worth noting that even after the Playa Azul earthquake, the seismic potential of the gap could not have been defined with any precision (HAVSKOV et al., 1983; LEFEVRE and MCNALLY, 1985).

Acknowledgements

We would like to thank many people who discussed with us over the years the issues of subduction mechanics. They include Kei Aki, Susan Beck, Doug Christensen, Hiroo Kanamori, Thorne Lay, Jim Rice, Larry Ruff, Susan Schwartz, Bill Spence, Bill Stuart and Bart Tichelaar. Also, we are grateful to Adam Dziewonski and Göran Ekström for many discussions and for smoothing our data retrieval efforts. Göran Ekström also provided his VBB inversion algorithms.

The research was supported primarily by the USGS, under grants 14–08–0001–G–1367 and 1788, and in part by NSF, under grant EAR–90–04556.

REFERENCES

ASTIZ, L. (1987), I. *Source Analysis of Large Earthquakes in Mexico*. II. *Study of Intermediate-depth Earthquakes and Interplate Seismic Coupling*, Ph.D. Thesis.

ASTIZ, L., and KANAMORI, H. (1986), *Interplate Coupling and Temporal Variation of Mechanisms of Intermediate-depth Earthquakes in Chile*, Bull. Seismol. Soc. Am. *76*, 1614–1622.

ASTIZ, L., LAY, T., and KANAMORI, H. (1988), *Large Intermediate Depth Earthquakes and the Subduction Process*, Phys. Earth Planet. Inter. *53*, 80–166.

BURBACH, G. V., and FROHLICH, C. (1986), *Intermediate and Deep Seismicity and Lateral Structure of Subducted Lithosphere in the Circum-Pacific Region*, Rev. Geophys. *24*, 833–874.

CHRISTENSEN, D. H., and RUFF, L. J. (1983), *Outer-rise Earthquakes and Seismic Coupling*, Geophys. Res. Lett. *10*, 697–700.

CHRISTENSEN, D. H., and RUFF, L. J. (1988), *Seismic Coupling and Outer-rise Earthquakes*, J. Geophys. Res. *93*, 13421–13444.

COOPER, P., and TAYLOR, B. (1987), *Seismotectonics of New Guinea: A Model for Arc Reversal Following Arc-continent Collision*, Tectonics *6*, 53–67.

CURTIS, J. W. (1973), *The Spatial Seismicity of Papua New Guinea and the Solomon Islands*, J. Geol. Soc. Australia *20*, 1–20.

DMOWSKA, R., and LOVISON, L. C. (1988), *Intermediate-term Seismic Precursors for Some Coupled Subduction Zones*, Pure Appl. Geophys. *126*, 643–664.

DMOWSKA, R., RICE, J. R., LOVISON, L. C., and JOSELL, D. (1988), *Stress Transfer and Seismic Phenomena in Coupled Subduction Zones During the Earthquake Cycle*, J. Geophys. Res. *93*, 7869–7884.

DZIEWONSKI, A. M., CHOU, T. A., and WOODHOUSE, J. H. (1981), *Determination of Earthquake Source Parameters from Waveform Data for Studies of Global and Regional Seismicity*, J. Geophys. Res. *86*, 2825–2852.

EKSTRÖM, G. A. (1987), *A Broad Band Method of Earthquake Analysis*, Ph.D. Thesis, Harvard University, Cambridge, Massachusetts.

EKSTRÖM, G. A. (1989), *A Very Broad Band Inversion Method for the Recovery of Earthquake Source Parameters*. Tectonophysics. *166*, 73–100.

GUTENBERG, B., and RICHTER, C. F, *Seismicity of the Earth* (1949).

HAVSKOV, J., SINGH, S. K., NAVA, E., DOMINGUEZ, T., and RODRIGUEZ, M. (1983), *Playa Azul, Michoacan, Mexico, Earthquake of 25 October, 1981 ($M_s = 7.3$)*, Bull. Seismol. Soc. Am. *73*, 449–458.

JARRARD, R. D. (1986), *Relations Among Subduction Parameters*, Rev. Geophys. *24*, 217–284.

JOHNSON, T., and MOLNAR, P. (1972), *Focal Mechanisms and Plate Tectonics of the Southwest Pacific*, J. Geophys. Res. *77*, 5000–5032.

KANAMORI, H., *The nature of seismicity patterns before large earthquakes*, In *Earthquake Prediction* (eds. Simpson D. W., and Richards P. G.) (Am. Geophys. Union 1981) pp. 1–19.

KELLEHER, J., SYKES, L., and OLIVER, J. (1973), *Possible Criteria for Predicting Locations and their Applications to Major Plate Boundaries of the Pacific and the Caribbean*, J. Geophys. Res. *78*, 2547–2583.

LAY, T., ASTIZ, L., KANAMORI, H., and CHRISTENSEN, D. H. (1989), *Temporal Variation of Large Interplate Earthquakes in Coupled Subduction Zones*, Phys. Earth Planet. Inter. *54*, 258–312.

LAY, T., and KANAMORI, H. (1980), *Earthquake Doublets in the Solomon Islands*, Phys. Earth Planet. Inter. *21*, 283–304.

LeFEVRE, L. V., and McNALLY, K. C. (1985), *Stress Distributions and Subduction of Aseismic Ridges in the Middle America Subduction Zone*, J. Geophys. Res. *96*, 4495–4510.

LOVISON, L. C., DMOWSKA, R., and DUREK, J. (1988), *Mechanics of Subduction as Inferred from Earthquake Cycle Observations: New Britain Area*, Abstract U51-104, AGU Fall Meeting Program, p. 153.

McCANN, W. R., NISHENKO, S. P., SYKES, L. R., and KRAUSE, J. (1979), *Seismic Gaps and Plate Tectonics: Seismic Potential for Major Boundaries*, Pure Appl. Geophys. *117*, 1082–1147.

NISHENKO, S. P. (1991), *Circum-Pacific Seismic Potential: 1989–1999*, Pure Appl. Geophys. *135*, 169–259.

PASCAL, G. (1979), *Seismotectonics of the Papua New Guinea — Solomon Islands Region*, Tectonophysics *57*, 7–34.

SCHWARTZ, S. Y., LAY, T., and RUFF, L. J. (1989), *Source Process of the Great 1971 Solomon Islands Doublet*, Phys. Earth Planet. Inter. *56*, 294–310.

(Received April 26, 1991, accepted May 20, 1991)

PAGEOPH, Vol. 136, No. 4 (1991)

0033–4553/91/040479–20$1.50 + 0.20/0

Size of Earthquakes in Southern Mexico from Indirect Methods

José A. Canas[1]

Abstract — L_g and duration magnitude formulas, as well as seismic moment relations, are derived for southern Mexico. The derivation uses a method based on the coda-Q value for this region obtained in another study (CANAS *et al.*, 1988) that needs only one seismographic station.

The results obtained are:

$$m_{L_g} = 2.37(\pm 0.16) + 1.00 \log \Delta + \log \frac{A^*}{T} \quad \text{for} \quad \Delta \le 400 \text{ km}, \quad m_{L_g} < 5.0.$$

A^* indicates reduced amplitudes.

$$m_{L_g} = -0.34(\pm 0.15) + 2.06 \log \Delta + \log \frac{A^*}{T} \quad \text{for} \quad 400 < \Delta \le 1200 \text{ km}, \quad m_{L_g} < 5.0$$

$$m_\tau = -1.48(\pm 0.32) + 2.31(\pm 0.15)\log \tau + 0.00042(\pm 0.00012)\Delta$$
$$\text{for} \quad 2.4 \le m_\tau < 5.0$$

$$m_\tau = -0.59(\pm 0.46) + 1.68(\pm 0.24)\log \tau + 0.00236(\pm 0.00049)\Delta$$
$$\text{for} \quad 2.4 \le m_\tau < 3.6$$

$$m_\tau = -0.53(\pm 0.40) + 1.96(\pm 0.18)\log \tau + 0.00024(\pm 0.00011)\Delta$$
$$\text{for} \quad 3.6 \le m_\tau < 5.0.$$

Using known seismic moments of some of the earthquakes in this study, it is found that the following relation satisfies the data:

$$\log M_0 = 15.84(\pm 0.82) + (1.40 \pm 0.21)m_\tau$$
$$\text{for} \quad 2.4 \le m_\tau < 5.0$$

where M_0 is the dyne-cm.

Several relations between L_g magnitude, duration magnitude, and seismic moments with respect to the ratio A_p of the L_g reduced amplitude to the L_g peak-to-peak short-period vertical component amplitudes have been found. These have the form:

$$m_{L_g} = 4.36(\pm 0.002) + (-1.15 \pm 0.03)\log A_p$$
$$\text{for} \quad 2.4 \le m_{L_g} < 5.0$$

$$m_\tau = 4.35(\pm 0.002) + (-1.12 \pm 0.03)\log A_p$$
$$\text{for} \quad 2.4 \le m_\tau < 5.0$$

$$\log M_0 = 21.91(\pm 0.02) + (-1.56 \pm 0.05)\log A_p$$
$$\text{for} \quad 2 \times 10^{19} \le M_0 \le 5 \times 10^{22}.$$

[1] Dep. Ingeniería del Terreno y Cartográfica. ETSICCP, Universidad Politécnica de Cataluña. Barcelona, Spain.

The average excitation factor obtained for all the data used in this study is:

$$B(f_p) = (1.12 \pm 0.98) \times 10^{-26} \text{ cm sec/dyne-cm.}$$

This value agrees with a relation in which the coda excitation factor is proportional to Q^{-1} as found by others.

A new method to determine magnitudes and seismic moments is introduced in this paper. It may be very useful because these parameters can be determined using only a single variable.

Key words: Magnitudes, seismic moment, coda-Q.

Introduction

The L_g phase and the duration of local and regional earthquakes have been used widely to determine local and regional magnitude formulas (e.g., NUTTLI, 1973; REAL and TENG, 1973; HERRMANN, 1975; BOLLINGER, 1979; SUTEAU and WHITCOMB, 1979; NUTTLI, 1980; HAVSKOV and MACIAS, 1983; BAKUN, 1984a, 1984b). The use of duration is especially important when there is a lack of appropriate instruments to estimate magnitude from amplitude data. The Seismological Service of Mexico calculates the M_L magnitudes using a Wood-Anderson instrument in Mexico D. F. Due to the low amplification of this seismograph no estimates of magnitudes are given for numerous earthquakes that are recorded at other stations in southern Mexico. The L_g phase is especially important for obtaining magnitude formulas when the first P arrival is too small or difficult to determine, basically for small earthquakes located at regional distances.

Determination of seismic moments is important for the assessment of magnitudes (e.g., BAKUN and LINDH, 1977; HERRMANN, 1980; DWYER *et al.*, 1983; BAKUN, 1984a) and for tectonic implications (e.g., BRUNE, 1970; REYES *et al.*, 1979; VALDÉS *et al.*, 1982; SINGH and HERRMANN, 1983).

Coda-Q values seem to be lower in the tectonically active provinces than in the stable ones (e.g., SINGH and HERRMANN, 1983; RAOOF, 1984; CANAS *et al.*, 1987). Since empirical relations between the coda excitation factors and Q have been found (e.g., AKI, 1980; SINGH and HERRMANN, 1983), determination of coda excitation factors seems to be promising to determine tectonic behaviour.

The goals of this study are to determine magnitude formulas from the L_g phase and duration, seismic moment formulas and the average coda excitation factor for southern Mexico. A relationship between m_{L_g}, m_τ (τ = duration), M_0 and the ratio of the L_g reduced amplitude (HERRMANN, 1980) to the L_g peak-to-peak amplitude will be also determined.

The seismographic station used in the above determinations belongs to the Seismological Network of the National Seismological Service of the Geophysical Institute of Mexico. It has the same characteristics as those of the WWSSN system, and the amplification curves and the calibration constants are properly known. This study uses the records of the short-period vertical component seismograph.

Data

All earthquakes (Table 1, Figure 1) have been located by the National Seismological Survey. Most of the seismicity in southern Mexico occurs along the Pacific coast. Therefore, most of the earthquakes used in this study have been chosen from the region extending from the Pacific coast of Guatemala to the coast of the Jalisco state in Mexico. The earthquakes, then, sample a wide geographic region which helps to average out problems due to differences in source radiation and crustal Q structure.

The vertical component of the short-period seismograms of the OXM seismographic station (Figure 1) has been analyzed to obtain L_g and duration magnitude formulas. The duration of the earthquakes, in this study, is the time lapsed between the onset of the first arrival and the point where the signal falls below the noise level. The durations and amplitude data have been read by the same person (the author) in order to eliminate any possible human errors due to different criteria in assessing the duration and amplitudes. Data for L_g magnitude studies are the same as used by CANAS et al. (1988) to determine the attenuation coefficient for the L_g sustained amplitude on the short-period vertical component of the OXM seismographic station.

Method

a) L_g Magnitude Formulas

The peak-to-peak L_g reduced amplitude data and the $\gamma - L_g$ attenuation coefficient determined by CANAS et al. (1988) for southern Mexico are used in this study. They found that for the epicentral distances from about 30 km–1200 km, the $\gamma - L_g$ attenuation was 0.00142 ± 0.00015 km^{-1} at a period of 1 second. The theoretical attenuation was 0.00142 ± 0.00015 km^{-1} at a period of 1 second. The theoretical attenuation curve for this γ-value can be approximated by two straight lines as shown in Figure 2. These lines intersect at a distance of about 400 km, therefore two L_g magnitude formulas will be derived. The first one is for distances ≤ 400 km, and the second one for distances between 400 and 1200 km.

Most local magnitude formulas used in Mexico are based on California parameters. Since the $\gamma - L_g$ attenuation is about three times larger in California than in southern Mexico (CANAS et al., 1988) it is unlikely that the adapted California magnitude formulas are the best ones for southern Mexico. VALDÉS et al. (1982) found this problem when they tried to compare local magnitude formulas based on coda length of local magnitudes that were determined assuming a California range amplitude relation.

The procedure to obtain the L_g magnitude formulas is the usual one, and it will be discussed in the next section.

Table 1

Earthquakes recorded at the OXM Seismographic Station used in this study

Date	H_0	Location		h	m_b	$m_{L_g}^{(*)}$	$m_{\tau}^{(*)}$	$M_o^{(*)}$ dyn-cm
6/Oct/78	19:46:46.0	14.1N,	91.5W	92	4.5	4.42	4.35	$8.51*10^{21}$
24/Oct/78	21:24:33.0	13.7N,	93.09W	33		4.45	4.42	$1.03*10^{22}$
26/Oct/78	06:00:26.0	18.7N,	106.13W	47	4.5	4.15	4.26	$6.37*10^{21}$
26/Oct/78	09:55:51.4	14.99N,	93.38W	2	4.0	4.09	4.30	$7.24*10^{21}$
16/Nov/78	15:55:53.0	18.0N,	95.08W	33	4.7	4.60	4.28	$6.58*10^{21}$
30/Nov/78	05:10:01.0	15.9N,	96.6W	33		4.1	4.5	$1.34*10^{22}$
30/Nov/78	13:15:24.0	15.4N,	96.44W	50	4.1	4.5	4.49	$1.20*10^{22}$
30/Nov/78	23:54:09.0	15.5N,	96.4W	33		4.65	4.40	$9.68*10^{21}$
2/Dec/78	20:27:41.9	16.07N,	96.39W	50	4.8	4.56	4.49	$1.17*10^{22}$
2/Dec/78	23:34:30.0	15.7N,	96.45W	57	4.7	4.66	4.49	$1.34*10^{22}$
3/Dec/78	17:11:12.0	15.7N,	97.4W	33	4.4	4.47	4.38	$9.38*10^{21}$
5/Dec/78	13:01:38.0	16.2N,	96.5W	51	4.1	4.13	4.22	$5.60*10^{21}$
5/Dec/78	19:21:33.0	15.3N,	96.4W	33	4.1	4.38	4.38	$9.38*10^{21}$
11/Dec/78	05:06:13.0	15.8N,	100.2W	33	4.5	4.41	4.34	$7.98*10^{21}$
11/Dec/78	11:52:16.0	16.8N,	99.4W	95	4.3	4.29	4.46	$1.21*10^{22}$
11/Dec/78	15:28:43.0	15.7N,	96.69W	19	4.4	4.54	4.49	$1.29*10^{22}$
13/Dec/78	09:48:59.0	15.9N,	98.5W	33	4.2	4.36	4.45	$1.17*10^{22}$
18/Dec/78	21:50:04.0	15.7N,	96.55W	52	4.5	4.96	4.46	$1.17*10^{22}$
22/Dec/78	04:54:28.0	16.94N,	100.07W	71	4.3	4.40	4.60	$1.91*10^{22}$
28/Dec/78	08:52:52.8	16.07N,	96.43W	42	4.6	4.79	4.70	$2.63*10^{22}$
6/Jan/79	09:21:55.0	15.3N,	93.88W	122	4.6	4.65	4.63	$2.10*10^{22}$
6/Jan/79	20:21:29.7	19.33N,	99.17W			2.39	2.51	$2.25*10^{19}$
7/Jan/79	11:48:33.8	19.45N,	98.82W	42		3.06	3.10	$1.47*10^{20}$
10/Jan/79	21:28:21.1	16.30N,	94.32W	104	4.9	4.71	4.49	$1.29*10^{22}$
12/Jan/79	23:17:10.1	16.96N,	93.72W	165	4.9	4.9	4.62	$2.03*10^{22}$
14/Jan/79	21:41:43.0	14.5N,	92.0W	76	4.7	4.4	4.46	$1.17*10^{22}$
31/Jan/79	10:14:58.3	16.49N,	96.65W	51	4.8	4.68	4.72	$2.81*10^{22}$
5/Feb/79	17:57:06.0	14.1N,	93.07W	43	4.8	4.29	4.62	$2.03*10^{22}$
13/Feb/79	02:37:07.9	16.78N,	94.56W	117	4.7	4.65	4.50	$1.38*10^{22}$
24/Feb/79	17:49:02.4	17.14N,	94.00W	192	4.6	4.69	4.51	$1.43*10^{22}$
2/Mar/79	12:29:30.0	14.2N,	92.94W	52	4.3	4.30	4.53	$1.47*10^{22}$
22/Feb/79	19:44:52.1	19.91N,	100.35W	2		3.58	3.50	$4.83*10^{20}$
22/Feb/79	20:23:31.3	19.93N,	100.21W	0		2.80	2.79	$5.75*10^{19}$
23/Feb/79	06:29:53.3	19.86N,	100.26W	2		2.87	2.90	$7.94*10^{19}$
23/Feb/79	16:25:21.0	19.93N,	100.29W	4		2.79	2.71	$4.31*10^{19}$
1/Mar/79	04:45:22.1	19.92N,	100.27W	4		3.51	3.29	$2.54*10^{20}$
4/Jun/80	14:29:44.7	15.79N,	95.70W	33		4.43	4.19	$5.08*10^{21}$
7/Jun/80	08:55:26.8	19.19N,	99.92W	33		2.48	2.61	$3.12*10^{19}$
7/Jun/80	09:46:07.8	19.29N,	100.17W	33		2.76	2.67	$3.91*10^{19}$
7/Jun/80	12:29:59.3	15.66N,	98.12W	33		3.76	3.84	$1.81*10^{21}$
9/Jun/80	06:24:44.3	16.11N,	98.52W	33		4.10	4.13	$4.33*10^{21}$
11/Jun/80	02:38:25.0	13.62N,	91.22W	33		4.88	4.85	$4.13*10^{22}$
14/Jun/80	12:47:10.2	15.74N,	97.44W	33		4.36	4.38	$9.68*10^{21}$
16/Jun/80	20:36:12.2	16.87N,	100.51W	15		3.79	3.77	$1.40*10^{21}$
24/Jun/80	08:51:06.4	17.42N,	100.92W	33		3.61	3.63	$6.46*10^{20}$
29/Jun/80	16:31:34.7	17.82N,	99.72W	33		3.47	3.46	$4.68*10^{20}$
2/Oct/80	14:07:41.0	17.33N,	101.96W	33		3.78	3.76	$1.40*10^{21}$

Table 1 (*Contd*)

Date	H_0	Location		h	m_b	$m_{L_g}^{(*)}$	$m_\tau^{(*)}$	$M_o^{(*)}$ dyn-cm
5/Oct/80	01:59:55.0	17.14N,	101.82W	33		3.99	4.07	$3.80*10^{21}$
6/Oct/80	07:06:42.0	15.83N,	97.32W	25		3.90	4.03	$3.13*10^{21}$
6/Oct/80	07:27:34.0	15.66N,	97.31W	25		3.93	4.01	$3.13*10^{21}$
9/Oct/80	12:36:08.1	16.94N,	95.30W	70		4.69	4.67	$2.47*10^{22}$
10/Oct/80	21:59:29.5	16.40N,	98.50W	33		4.03	4.02	$3.03*10^{21}$
11/Oct/80	06:07:37.0	15.18N,	97.05W	33		4.34	4.11	$4.06*10^{21}$
16/Oct/80	14:06:40.0	14.70N,	93.40W	33		4.85	4.73	$2.90*10^{22}$
16/Oct/80	14:51:22.5	14.17N,	93.09W	33	4.5	4.67	4.67	$2.39*10^{22}$
17/Oct/80	22:22:54.5	15.55N,	96.75W	33		4.39	4.20	$5.42*10^{21}$
18/Oct/80	22:46:58.0	18.77N,	105.42W	33		4.64	4.39	$1.00*10^{22}$
20/Oct/80	23:36:09.0	18.49N,	103.49W	33		4.0	4.05	$3.56*10^{21}$
24/Oct/80	16:25:51.0	17.92N,	98.49W	50		3.61	3.62	$9.51*10^{20}$
24/Oct/80	22:17:04.5	15.50N,	98.42W	50		3.63	3.91	$2.27*10^{21}$
24/Oct/80	23:00:20.0	17.82N,	98.23W	50		3.95	3.97	$2.75*10^{21}$
25/Oct/80	09:54:41.0	17.88N,	98.12W	50		3.59	3.63	$8.09*10^{20}$
28/Oct/80	09:44:09.6	15.94N,	96.44W	33		4.22	4.14	$4.47*10^{21}$
28/Oct/80	10:40:25.0	17.93N,	98.07W	52		3.57	3.79	$1.40*10^{21}$
3/Nov/80	23:11:38.0	16.71N,	95.24W	33		4.21	4.03	$3.24*10^{21}$
5/Nov/80	09:52:52.5	17.23N,	101.62W	33		3.82	3.72	$1.19*10^{21}$
22/Nov/80	00:18:46.0	14.84N,	93.79W	33		4.6	4.58	$1.79*10^{22}$
22/Nov/80	12:40:39.0	17.56N,	101.64W	33		3.35	3.43	$4.39*10^{20}$
30/Nov/80	04:46:31.5	17.99N,	101.26W	33		3.57	3.32	$3.08*10^{20}$
3/Dec/80	08:18:49.5	17.37N,	102.65W	33		3.99	4.10	$3.93*10^{21}$
3/Dec/80	17:02:35.0	17.89N,	98.21W	50		3.57	3.39	$3.73*10^{20}$
9/Dec/80	11:14:21.0	18.46N,	104.76W	33		4.24	4.28	$7.02*10^{21}$
13/Dec/80	12:40:30.0	67.13N,	101.59W	33		3.79	3.70	$1.12*10^{21}$
16/Dec/80	09:25:49.0	15.42N,	94.83W	33		4.44	4.39	$9.68*10^{21}$
18/Dec/80	10:45:07.0	15.07N,	98.28W	33		4.42	4.41	$1.03*10^{22}$
18/Dec/80	17:01:48.5	15.28N,	97.95W	33		4.31	4.41	$2.00*10^{21}$
20/Dec/80	00:10:41.0	15.27N,	98.98W	33		4.21	4.28	$6.80*10^{21}$
20/Dec/80	02:44:28.0	15.51N,	98.66W	33		3.89	3.83	$1.59*10^{21}$
20/Dec/80	09:05:39.0	14.86N,	98.71W	33		4.39	4.30	$7.48*10^{21}$
20/Dec/80	09:16:54.0	15.30N,	98.98W	33		3.98	4.08	$3.80*10^{21}$
21/Dec/80	01:43:04.5	18.12N,	102.38W	33		3.87	3.95	$2.58*10^{21}$
26/Dec/80	20:29:04.5	15.79N,	99.43W	33		4.08	3.95	$2.59*10^{21}$
26/Dec/80	20:35:35.0	16.16N,	99.78W	33		3.77	3.96	$2.58*10^{21}$
28/Dec/80	20:39:40.5	16.34N,	99.96W	33		3.79	3.70	$1.12*10^{21}$
29/Dec/80	08:58:02.0	16.57N,	98.08W	33		4.03	4.14	$4.46*10^{21}$
30/Dec/80	15:55:58.5	19.34N,	107.12W	33	4.8	4.52	4.83	$4.00*10^{22}$

(*)Calculated in this study.

b) *Duration Formulas*

Duration magnitude formulas, m_τ, have the general form (e.g., LEE et al., 1972; REAL and TENG, 1973):

$$m_\tau = a_0 + a_1 \log \tau + a_2 (\log \tau)^2 + a_3 \Delta \tag{1}$$

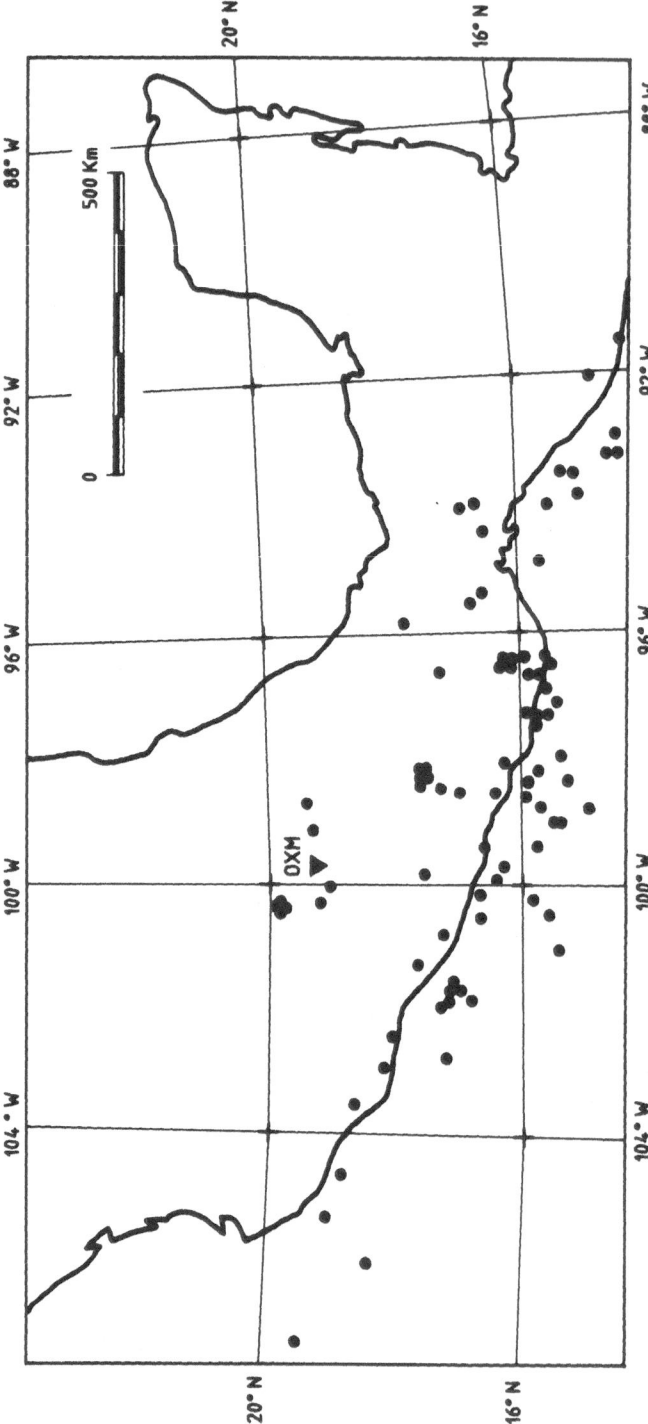

Figure 1

Location of the epicenters and the seismographic stations used in this study.

where a_0, a_1, a_2, a_3 are real constants to be determined from the data. τ is the duration of the seismogram in seconds and Δ is the epicentral distance in km. There are several approaches to determine how many terms in expression (1) must be used.

In this case, to determine the number of terms necessary in expression (1), a F-test for adding a new one is performed. Several starting formulas have been tried to detect the best expression, as for example the following:

$$m_\tau = a_0 + a_1 \log \tau$$

$$m_\tau = a_0 + a_2 (\log \tau)^2. \tag{2}$$

$$m\tau = a_0 + a_3 \Delta$$

F-test has been performed going from two to three terms, and from two to four terms. The best result at the 99% confidence level occurs when a_2 in expression (1) is taken to be zero, therefore the formula:

$$m_\tau = a_0 + a_1 \log \tau + a_3 \Delta \tag{3}$$

is the most appropriate one to describe the duration data for any different m_{L_g} magnitude range in this study. The details of the calculation of the coefficients and the number of formulas are given in the next section.

Magnitude Formula. Results

a) L_g Formula

L_g magnitude formulas can be represented as in NUTTLI (1973)

$$m_{L_g} = B + C \log \Delta + \log \frac{A^*}{T}. \tag{4}$$

C is obtained for each case from the slopes of the straight lines in Figure 2. C for distances equal or less than 400 km is equal to 1.00, and for the distance range $400 < \Delta \leq 1200$ km is 2.06. B is considered constant for the magnitude range in this study ($m_b < 5$) and it is obtained, in each case, using earthquakes with given magnitudes by the International Seismological Center (I.S.C.). T is the L_g wave period between 0.7 and 1.3 sec. The earthquakes used to calculate the constant C for distances ≤ 400 km have m_b magnitudes between 4.1 and 4.8 (Table 1). For the range $400 < \Delta \leq 1200$ km they extend from $m_b = 4.0$ to $m_b = 4.9$ (Table 1). Unfortunately no estimates of magnitude less than about 4.0 have been found in the I.S.C. Bulletin. Therefore, strictly speaking, the derived formulas must be used in the range between about 4.0 and 5.0; however, we will use the obtained formulas to make magnitude estimates for all the earthquakes in this study. We expect that

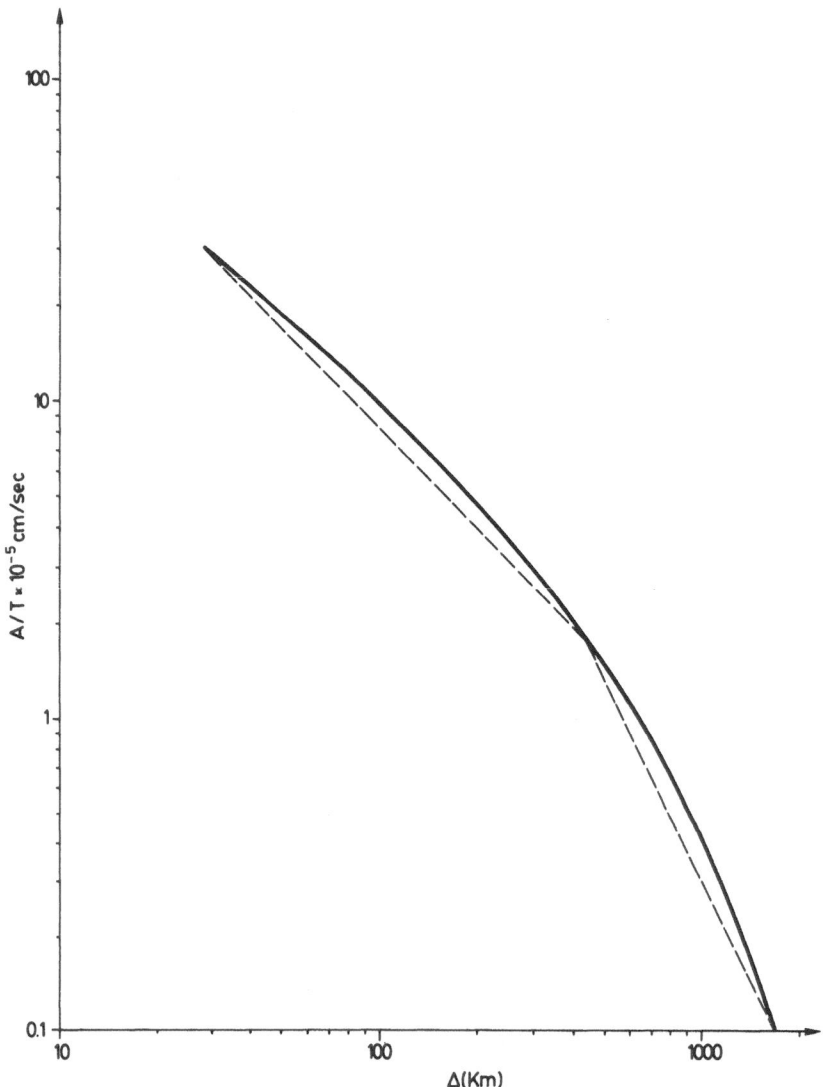

Figure 2

$\gamma - L_g$ attenuation curve obtained by CANAS *et al.* (1988) for southern Mexico. The γ value is 0.00142 (± 0.00015) km^{-1}. The two straight lines intersecting at 400 km represent one approach to the curve.

possible errors in the estimated magnitude will not be large. If teleseismic magnitude data for events less than about 4 become available, the derived formulas can be tested and corrected for low magnitudes, if necessary. A^* means the moment corrected amplitude as defined by HERRMANN (1980) and T is about 1 second in this study. The A data are the same used by CANAS *et al.* (1988) in the same region of this study.

The formulas obtained are:

$$m_{L_g} = 2.37(\pm 0.16) + 1.00 \log \Delta + \log \frac{A^*}{T}$$

$$\text{for} \quad \Delta \le 400 \text{ km} \tag{5}$$

$$m_{L_g} = -0.34(\pm 0.15) + 2.06 \log \Delta + \log \frac{A^*}{T}$$

$$\text{for} \quad 400 \le \Delta \le 1200 \text{ km.} \tag{6}$$

The method described above to obtain magnitude formulas is useful for regions where just one seismographic station is available, and no other sources provide confident amplitude data. Table 1 lists the m_{L_g} magnitudes obtained by expressions (5) and (6) for all the earthquakes in this study.

b) *Duration Formula*

In Figure 3, the m_{L_g} magnitudes of the earthquakes in Table 1 are plotted versus $\log \tau$, τ being the duration in seconds. It appears (Figure 3) that the slope of the

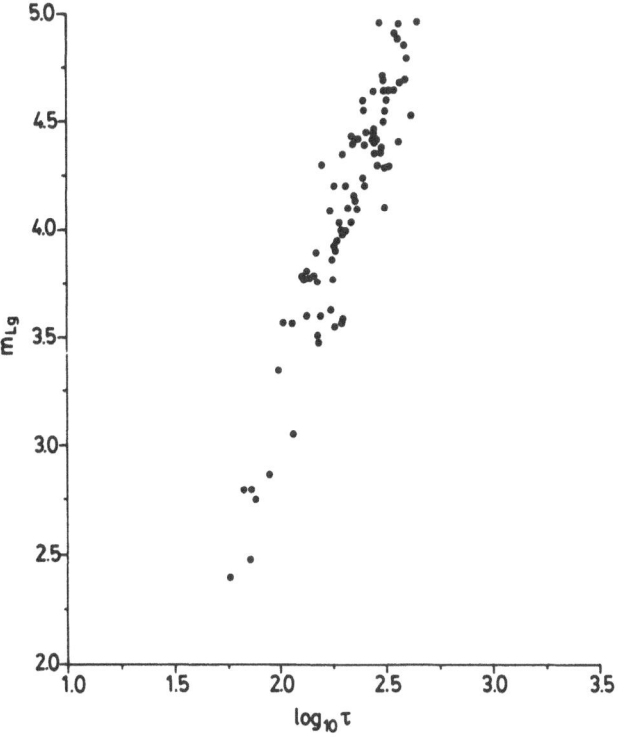

Figure 3

L_g magnitude (m_{L_g}), determined in this study, versus logarithm of duration (τ) in seconds.

trend is changing slightly at about $m_{L_g} = 3.6$. To take into account this fact, three duration magnitude formulas have been tested. The first one is for m_{L_g} magnitudes between 2.4 and 3.6, and the last one for magnitudes between 3.6 and 5.0. In all cases, as stated in the method section, the best formulas are given by expression (3).

Regression analysis applied to expression (3) using the m_{L_g} magnitude data for $2.4 \leq m_{L_g} \leq 5.0$, $2.4 \leq m_{L_g} \leq 3.6$ and $3.6 \leq m_{L_g} \leq 5.0$ give the following results:

$$m_\tau = -1.48(\pm 0.32) + (2.31 \pm 0.15) \log \tau + (0.00042 \pm 0.00012)\Delta$$

$$\text{for} \quad 2.4 \leq m_\tau \leq 5.0 \tag{7}$$

$$\text{R.M.S.} = 0.19 \quad \text{C.C.} = 0.93.$$

R.M.S. and C.C. indicate root-mean-square and correlation coefficients, respectively.

$$m_\tau = -0.59(\pm 0.46) + (1.68 \pm 0.24) \log \tau + (0.00236 \pm 0.00049)\Delta$$

$$\text{for} \quad 2.4 \leq m_\tau < 3.6 \tag{8}$$

$$\text{R.M.S.} = 0.13 \quad \text{C.C.} = 0.95$$

$$m_\tau = -0.53(\pm 0.40) + (1.96 \pm 0.18) \log \tau + (0.00024 \pm 0.00011)\Delta$$

$$\text{for} \quad 3.6 \leq m_\tau < 5.0 \tag{9}$$

$$\text{R.M.S.} = 0.17 \quad \text{C.C.} = 0.95.$$

From the above results it seems that expressions (8) and (9) give a slightly better fit to the data set than expression (7). Therefore, expressions (8) and (9) are the ones to be used to determine magnitude from duration.

The coefficients of the epicentral distance terms in expressions (8) and (9) are quite different. In fact, the coefficient in formula (8) is about ten times greater than the one in expression (9). One explanation may come from the fact that most of the earthquakes in Figure 1, with magnitude lower than 3.6, are located in the Central Volcanic Axis of Mexico. CANAS (1986) found that Q values at 1 Hz for earthquakes in the Volcanic Axis and recorded at the OXM seismographic station are about 130. Since the average 1 Hz Q value for southern Mexico is about 490, the epicentral coefficients in expressions (8) and (9) seem to be reasonably explained. Expression (7) may be considered as an average for the whole magnitude range. In fact, HAVSKOV and MACIAS (1983) using 61 events with similar distribution as those in this study, and 12 short-period seismographic stations, 11 of them belonging to the Engineering Institute of Mexico, found the following formula:

$$m_\tau = -1.59 + 2.40 \log \tau + 0.00046 \, \Delta. \tag{10}$$

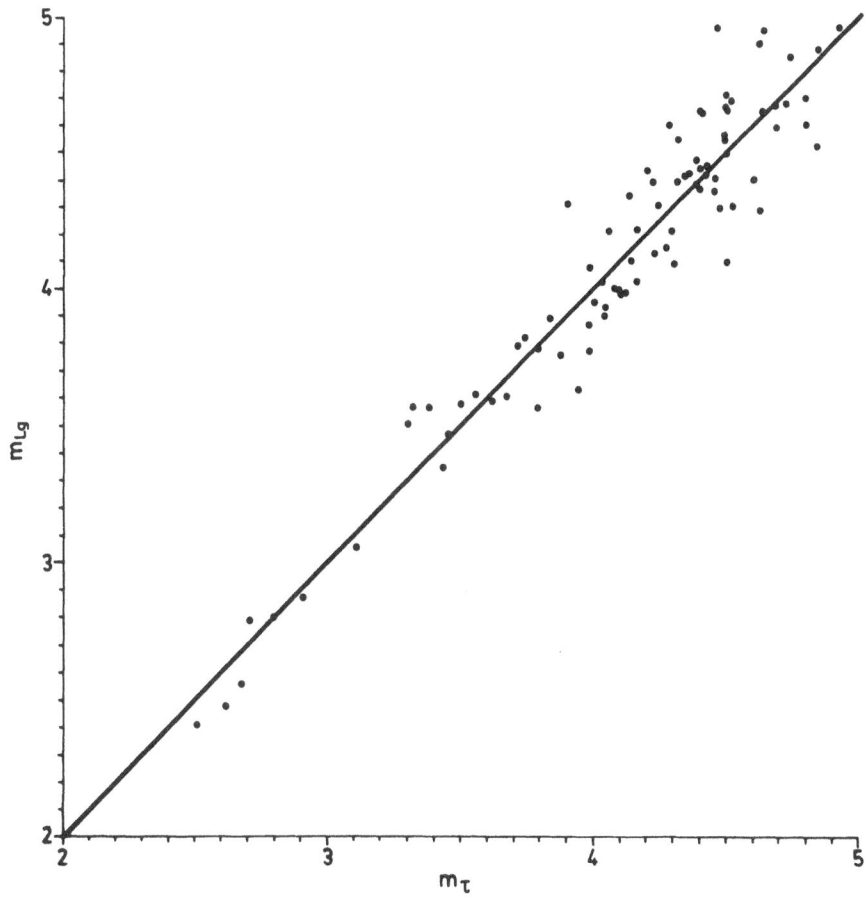

Figure 4

L_g magnitude (m_{L_g}) versus duration magnitude (m_τ) obtained in this study. The straight line has been obtained using a linear least-squares fit.

Although it is difficult to make any comparison between formulas (7) and (10), due to the different instruments used, they give a similar coefficient, partially supporting the results in this study. HAVSKOV and MACIAS (1983) used events in the magnitude range 4.0–5.8. Therefore the expression found in this work tentatively can be extended up to magnitudes of about 6. Table 1 lists the duration magnitudes obtained using expressions (8) and (9). Figure 4 presents the m_{L_g} magnitude versus m_τ magnitudes. The correlation between both is excellent and the resulting formula is:

$$m_{L_g} = (-0.018 \pm 0.133) + (1.008 \pm 0.032)m_\tau \qquad (11)$$

$$\text{C.C.} = 0.96.$$

Magnitude and Seismic Moment Relations

Using 9 earthquakes (Table 2) with calculated seismic moments (VALDÉS et al., 1982) a correlation between seismic moments and duration magnitude has been found. The earthquakes have been selected such that the quality, as given by VALDÉS et al. (1982), was the best one, and also that the duration at the OXM station could be properly defined. The magnitude range of those earthquakes extends from about 3 to about 5. Therefore, the resulting formula will be used to assess the magnitudes of all the earthquakes. Figure 5 presents the correlation between $\log M_0$, M_0 being the seismic moment in dyn-cm, and the duration magnitude m_τ. The resulting formula is:

$$\log M_0 = 15.84(\pm 0.82) + (1.40 \pm 0.21)m_\tau. \tag{12}$$

Expression (12) compares well with a similar relation found by VALDÉS et al. (1982) working in a small region of the Guerrero State (Mexico). They correlated seismic moments and local magnitudes using California formulas. The M_0 values given by expression (12) and by the VALDÉS et al. (1982) formula for the magnitude range $2.4 \le m_\tau \le 5.0$ do not differ more than about one half of the unit, therefore the local formula of VALDÉS et al. (1982) seems to be in good agreement with the more general formula obtained in this study.

Relations between Magnitudes, Seismic Moments and the Ratio of Amplitudes A_p

The seismic moment corrected amplitude (HERRMANN, 1980), A_c, is given by:

$$A_c = A_p \times A_{\text{peak}} \tag{13}$$

Table 2

Earthquakes used to obtain the relationship between M_0 and m_τ in this study

Date	Ho	Location		h	m_τ^*	$M_0^{(**)}$ dyn-cm
16/Mar/79	06:04:53.38	17.4N,	101.4W	26	4.71	$12.7 * 10^{21}$
16/Mar/79	10:10:30.94	17.3N,	101.3W	25	4.44	$8.5 * 10^{21}$
17/Mar/79	08:25:56.46	17.4N,	101.2W	21	3.40	$1.4 * 10^{21}$
17/Mar/79	13:47:44.03	17.5N,	101.2W	26	3.23	$2.8 * 10^{20}$
17/Mar/79	15:50:32.62	17.5N,	101.1W	27	3.22	$1.9 * 10^{20}$
17/Mar/79	18:57:33.03	17.4N,	101.4W	25	3.65	$1.1 * 10^{21}$
18/Mar/79	02:39:22.27	17.4N,	101.2W	20	3.51	$8.2 * 10^{20}$
20/Mar/79	00:27:51.66	17.3N,	101.4W	30	4.82	$16.7 * 10^{22}$
21/Mar/79	05:32:30.55	17.4N,	101.3W	29	2.99	$5.3 * 10^{19}$

(*)Calculated in this study.
(**)From VALDÉS et al. (1982).

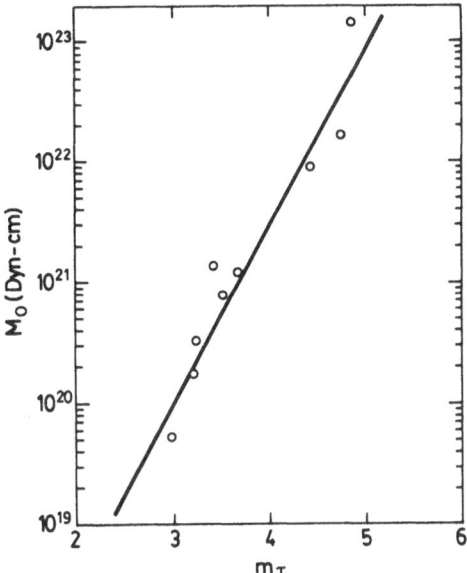

Figure 5

Obtained relationship between seismic moment (M_0) of VALDÉS *et al.* (1982) and duration magnitude (m_τ) determined in this study.

where A_p is the value that corresponds to 1 cm peak-to-peak on the observed seismograms once it is matched to the theoretical coda-shape curve. A_{peak} represents the L_g peak-to-peak maximum amplitude.

Expression (13) can be written as:

$$A_p = \frac{A_c}{A_{\text{peak}}}. \tag{14}$$

A_p data are obtained from CANAS *et al.* (1988).

Figures 6 to 8 present the results obtained after applying a linear least-squares fitting between magnitudes, seismic moments and A_p data; the results are:

$$m_{L_g} = (4.36 \pm 0.02) + (-1.15 \pm 0.03)\log A_p$$

$$2.4 \leq m_{L_g} \leq 5.0 \tag{15}$$

$$\text{C.C.} = 0.97$$

$$m_{L_g} = (4.35 \pm 0.002) + (-1.12 \pm 0.03)\log A_p$$

$$2.4 \leq m_{L_g} \leq 5.0 \tag{16}$$

$$\text{C.C.} = 0.97$$

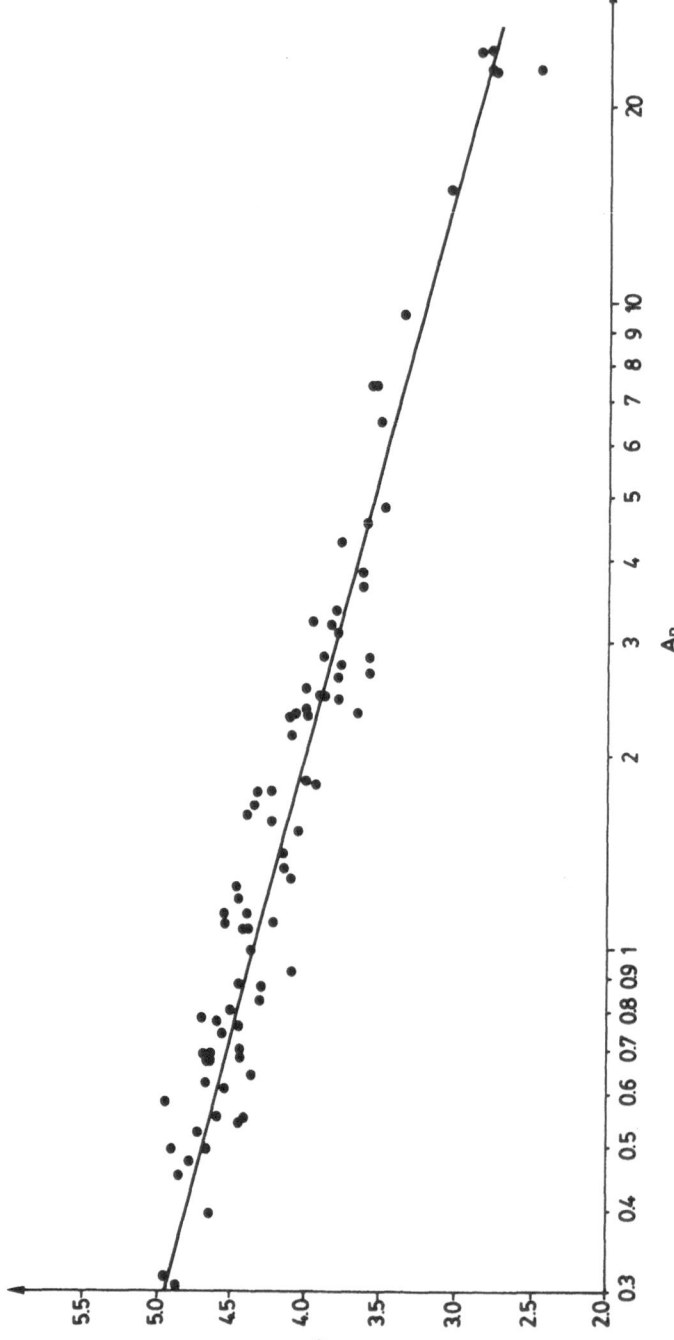

Figure 6

L_g magnitude versus log A_p values (see text for definition of A_p). The straight line indicates the linear least-squares fit to the data.

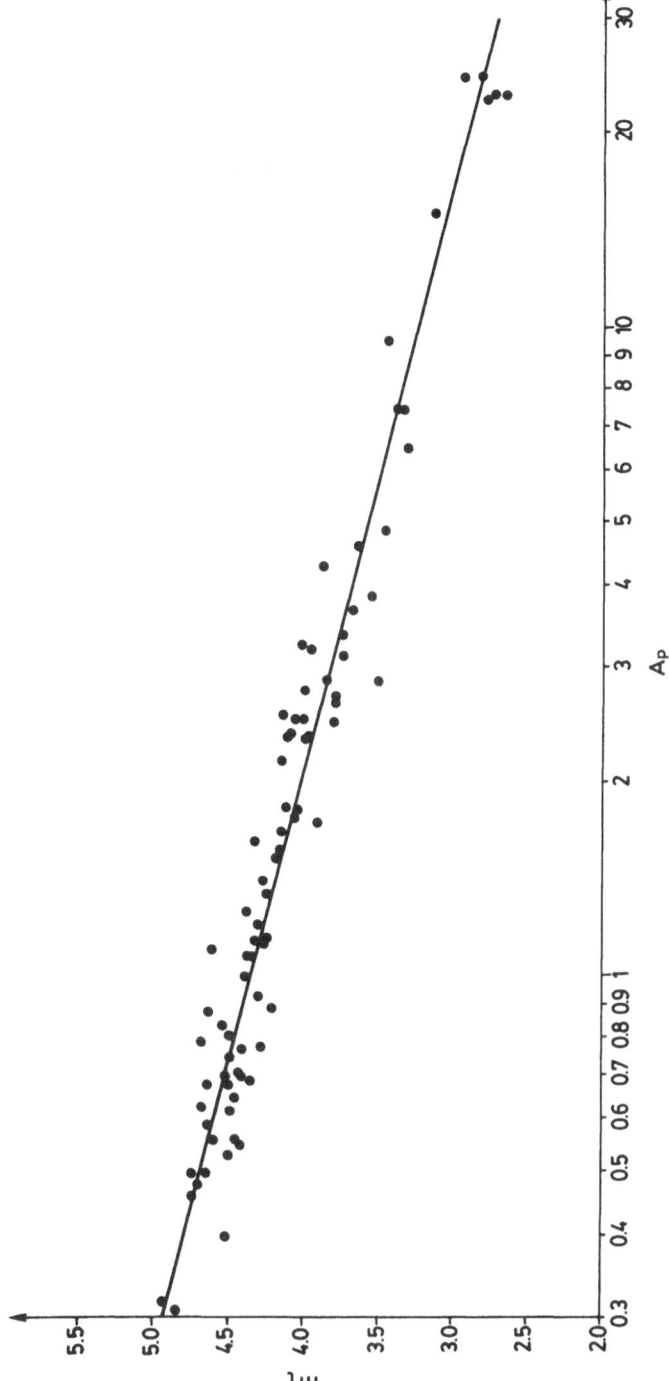

Figure 7

Duration magnitude versus log A_p values (see text for definition of A_p). The straight line indicates the linear least-squares fit to the data.

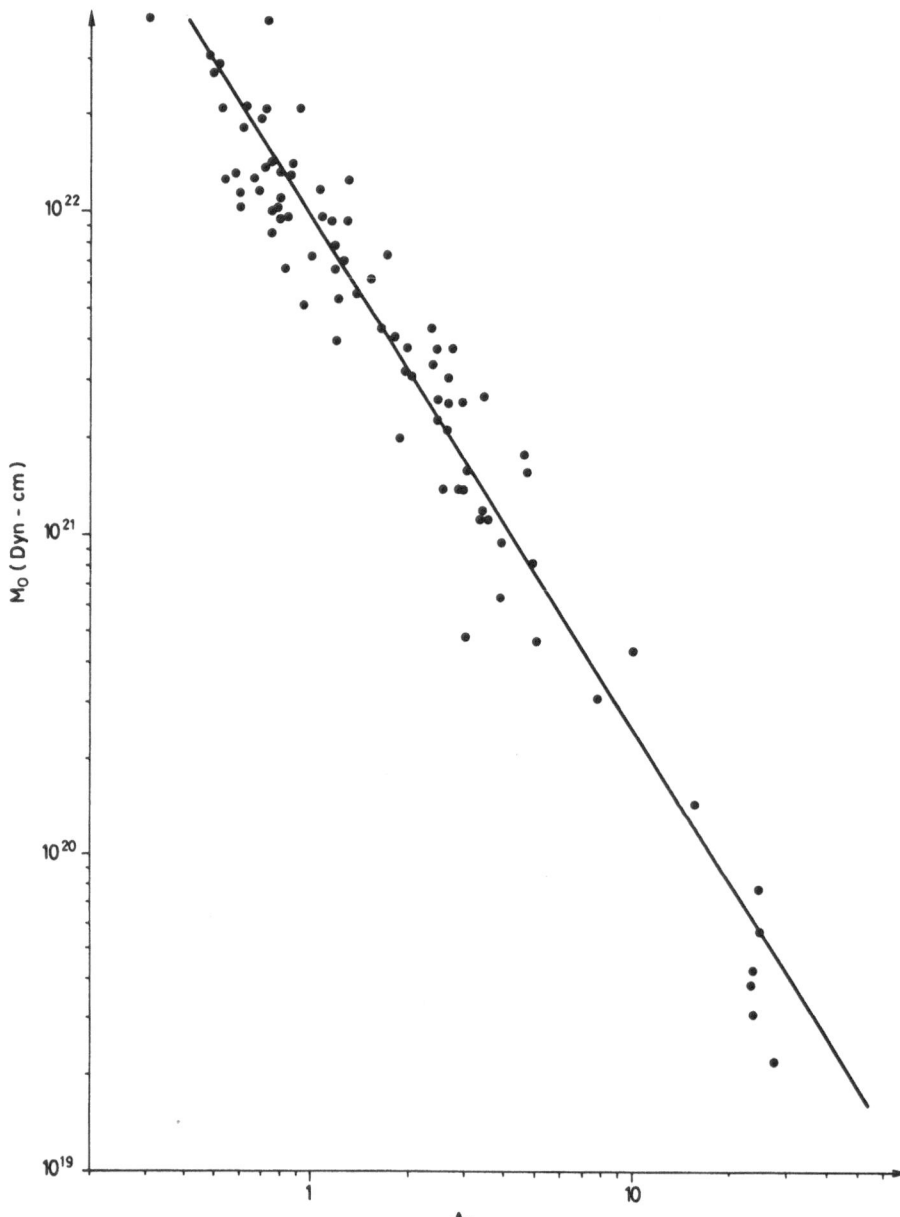

Figure 8
Logarithm of seismic moment (M_0) versus log A_p values (see text for definition of A_p). The straight line
indicates the linear least-squares fit to the data.

$$\log M_0 = (21.91 \pm 0.02) + (-1.56 \pm 0.05)\log A_p$$

$$2 \times 10^{19} \le M_0 \le 5 \times 10^{22} \tag{17}$$

$$\text{C.C.} = 0.96.$$

The matching between m_{L_g}, m_τ and $\log M_0$ with respect to $\log A_p$ is extremely good. The important conclusion from formulas (15) to (17) is that we are able to obtain L_g and duration magnitudes and seismic moments by just knowing a single number. A number that can be determined quickly and easily because the ratio in expression (14) is constant for the entire seismogram, therefore the readings of any phase amplitude and its corresponding seismic moment corrected amplitude provide the A_p value. Expression (14) must be viewed in a more general sense, since A_{peak} means really the peak-to-peak amplitude of any phase on the seismogram, and A_c the corresponding corrected amplitude.

Coda Excitation Factor

SINGH and HERRMANN (1983), using seismic moments of earthquakes in several regions of the United States, determined a relationship between the coda excitation factor, $B(f_p)$, where f_p is the predominant frequency, and the Q values. They obtained the relation:

$$\log B(f_p) = -22.355(\pm 0.748) - 1.076(\pm 0.273)\log Q. \tag{18}$$

This result suggests (SINGH and HERRMANN, 1983) that the coda excitation factor $B(f_p)$ varies as Q^{-1}, indicating in a qualitative sense that the more efficient is the medium as a wave transmitter, the less efficient it should be as a coda generator.

The factors $B(f_p)$ can be determined using the formula (HERRMANN, 1980):

$$\frac{A(t^*)}{\sqrt{B}} = Q^{-1/2}M_0 B(f_p)C(f_p, t^*) \tag{19}$$

where A is the peak-to-peak amplitude at time t^* ($= t/Q$), t being the time lapsed after the origin time, and $C(f_p, t^*)$ the coda shape function. $B(f_p)$ can be obtained if the values of A, Q, M_0 and C are known.

Values of A, Q and C are obtained from CANAS et al. (1988) and the values of M_0 are the ones determined in this study (Table 1). The average value obtained for southern Mexico by applying expression (19) is:

$$B(f_p) = (1.12 \pm 0.98) \times 10^{-36} \text{ cm sec/dyne-cm.} \tag{20}$$

The theoretical value determined from expression (18), and for an average Q of 489 (CANAS et al., 1988) is 5.64×10^{-26} cm/sec/dyne-cm. This value is consistent with the relation found by SINGH and HERRMANN (1983), indicating that the representative $B(f_p)$ value for southern Mexico agrees with a variation of $B(f_p)$ proportional to Q^{-1}.

Conclusions

L_g and duration magnitude formulas have been obtained for southern Mexico using the seismographic station OXM, located near Mexico D.F. L_g magnitude formulas are based on the $\gamma - L_g$ attenuation coefficient determined by CANAS et al. (1988) for the same region in this study. Duration magnitude formulas, m_τ, are determined by applying F-tests to several different magnitude formula approaches. Correlation between m_{L_g} and m_τ is extremely high. The average duration magnitude formula is practically the same obtained by HAVSKOV and MACIAS (1983), using twelve stations and a similar earthquake distribution. Although care must be taken in doing any comparison between both formulas, due to the different instruments used, the average duration formula seems to be consistent with that of HAVSKOV and MACIAS (1983).

Using 9 events with known seismic moments, a relationship between the seismic moments and duration magnitude has been found. This relation has been used to assess seismic moment estimates to all the earthquakes in this study.

Linear regression applied to L_g magnitudes, duration magnitudes and seismic moments versus the ratio of the seismic moment corrected amplitude with respect to the peak-to-peak L_g phase (A_p), provides a new and adequate way to obtain magnitudes and seismic moments for earthquakes in southern Mexico. Those relations are:

$$m_{L_g} = 4.36(\pm 0.20) + (-1.15 \pm 0.03)\log A_p$$

$$2.4 \leq m_{L_g} \leq 5.0$$

$$\text{C.C.} = 0.97$$

$$m_\tau = 4.35(\pm 0.02) + (-1.12 \pm 0.03)\log A_p$$

$$2.4 \leq m_\tau \leq 5.0$$

$$\text{C.C.} = 0.97$$

$$\log M_0 = 21.91(\pm 0.02) + (-1.56 \pm 0.05)\log A_p$$

$$2 \times 10^{19} \leq M_0 \leq 5 \times 10^{22}.$$

The method to obtain all the above relations used only one seismographic station. This can be useful when the amplitude data information is too detrimental to determine magnitudes, and when just a few stations are available in the region under study. Particularly, for southern Mexico, this approach has permitted the assessment of magnitudes and seismic moments for earthquakes with magnitudes less than about $m_{L_g} = 5.0$, when the Seismological Service of Mexico, for the period range of the studied earthquakes, was unable to assess magnitudes of most of the earthquakes with m_b magnitudes less than about 4.5.

To obtain magnitudes and seismic moments from A_p only a knowledge of Q for the region of interest is needed. The A_p values can be determined using any phase of the seismograms.

The average excitation factor of coda waves, $B(f_p)$, found in this study is $(1.12 \pm 0.98) \times 10^{-26}$ cm sec/dyn-cm. This value agrees very well with the empirical relation of SINGH and HERRMANN (1983), indicating that possibly $B(f_p)$ is proportional to Q^{-1} in southern Mexico.

The main goals of this work were to determine representative average magnitude and seismic moment formulas. Q lateral variations may be expected in southern Mexico. If so, a regionalization of the obtained formulas can be accomplished by employing the same method.

Comparing the epicentral distance coefficients in the m_τ formulas given by expressions (8) and (9), it is clear that the magnitude range $2.4 \le m_\tau < 3.6$ is larger than for the range $3.6 \le m_\tau \le 5.0$. Since the earthquakes with magnitudes less than 3.6 are located in or near the Mexican Volcanic Axis, it seems that the low-Q values associated with the Axis correlate well with the high epicentral distance coefficient in expression (8).

Acknowledgements

I thank the personnel of the National Seismological Service of Mexico for their assistance in selecting the seismograms used in this work, and Mrs. M. C. Rodriguez who kindly typed the paper.

This work has been partially supported by the Dirección General de Investigación Científica y Técnica (DGICYT), project number: PB87–0854.

REFERENCES

AKI, K. (1980), Scattering and Attenuation of Shear Waves in the Lithosphere, J. Geophys. Res. 85, 6496–6504.
BAKUN, W. H. (1984a), Seismic Moments, Local Magnitudes and Coda-duration Magnitudes for Earthquakes in Central California, Bull. Seism. Soc. Am. 74, 439–458.
BAKUN, W. H. (1984b), Magnitudes and Moments of Duration, Bull. Seism. Soc. Am. 74, 2335–2356.
BAKUN, W. H., and LINDH, A. G. (1977), Local Magnitudes, Seismic Moments, and Coda Durations for Earthquakes near Oroville, Bull. Seism. Soc. Am. 67, 615–629.
BOLLINGER, G. A. (1979), Attenuation of the L_g Phase and the Determination of m_b in the Southeastern United States, Bull. Seism. Soc. Am. 69, 45–63.

BRUNE, J. N. (1970), *Tectonic Stress and the Spectra of Seismic Shear Waves from Earthquakes*, J. Geophys. Res. *75*, 4997–5009.

CANAS, J. A. (1986), *Estudio del factor anelástico Q de la coda de los terremotos correspondientes a las regiones central y oriental del eje volcánico de México*, Geofísica Internacional *25*, 503–520.

CANAS, J. A., EGOZCUE, J. J., PUJADES, L., and PÉREZ, J. A. (1987), *Crustal Coda Q in the Iberian Peninsula*, Annales Geofísicae *5B*, 657–662.

CANAS, J. A., EGOZCUE, J. J., and PUJADES, L. (1988), *Seismic Attenuation in Southern Mexico Using the Coda-Q Method*, Bull. Seismol. Soc. Am. *78*, 1807–1817.

DWYER, J. J., HERRMANN, R. B., and NUTTLI, O. W. (1983), *Spatial Attenuation of L_g Wave in the Central United States*, Bull. Seismol. Soc. Am. *73*, 781–796.

HAVSKOV, J., and MACIAS, M. (1983), *A Coda-length Magnitude Scales for Mexico*, Geofísica Internacional *21*, 249–263.

HERRMANN, R. B. (1975), *The Use of Duration as a Measure of Seismic Moment and Magnitude*, Bull. Seismol. Soc. Am. *65*, 899–913.

HERRMANN, R. B. (1980), *Q Estimates Using the Coda of Local Earthquakes*, Bull. Seismol. Soc. Am. *70*, 447–468.

LEEM, W. H. K., BENNETT, R. E., and MEAGHAR, K. L. (1972), *A Method of Estimating Magnitude of Local Earthquakes from Signal Duration*, U.S. Geol. Survey, Open File Report, 28 pp.

NUTTLI, O. W. (1973), *Seismic Wave Attenuation and Magnitude Relations for Eastern North America*, J. Geophys. Res. *78*, 876–885.

NUTTLI, O. W. (1980), *The Excitation and Attenuation of Seismic Crustal Phases in Iran*, Bull. Seismol. Soc. Am. *70*, 469–485.

RAOOF, M. (1984), *Attenuation of High Frequency Earthquake Surface Waves in South America*, Master Thesis, Saint Louis University, Saint Louis, Mo.

REAL, C. R., and TENG, T. L. (1973), *Local Richter Magnitude and Total Signal Duration in Southern California*, Bull. Seismol. Soc. Am. *63*, 1809–1827.

REYES, A., BRUNE, J. N., and LOMNITZ, C. (1979), *Source Mechanism and Aftershock Study of the Colima, Mexico Earthquake of January 30, 1973*, Bull. Seismol. Soc. Am. *69*, 1819–1840.

SINGH, S., and HERRMANN, R. B. (1983), *Regionalization of Crustal Q in the Continental United States*, J. Geophys. Res. *81*, 527–538.

SUTEAU, A. M., and WHITCOMB, J. H. (1979), *A Local Earthquake Coda Magnitude and its Relation to Duration, Moment M_0 and Local Richter Magnitude. M_L*, Bull. Seismol. Soc. Am. *69*, 353–368.

VALDÉS, C., MEYER, R. P., ZÚÑIGA, R., HAVSKOV, J., and SINGH, S. K. (1982), *Analysis of Petatlan Aftershocks Having Coda Lengths Greater than 60 Seconds*, J. Geophys. Res. *87*, 8519–8527.

(Received November 29, 1990, revised/accepted May 21, 1991)

PAGEOPH, Vol. 136, No. 4 (1991)

0033–4553/91/040499–16$1.50 + 0.20/0

Numerical Simulation of the Earthquake Generation Process

Mircea Radulian,[1] Cezar-Ioan Trifu,[1] and Florin Octavian Cărbunar[2]

Abstract — A numerical algorithm is proposed for the simulation of the earthquake process during a seismic cycle. The algorithm is based on a heterogeneous discrete model of the fault plane and assumes there are two kinds of seismicity: background crack-like earthquakes and asperity-like events. An active zone of the fault contains an asperity distribution with a characteristic elementary area. The background seismicity randomly develops shear stress-free surfaces which tend to surround the asperities as in a 2D percolation process. The model parameters are taken from observations on the Vrancea (Romania) intermediate depth seismic region. The results emphasize the significant role of the geometry in the mechanism of the seismic failure. The algorithm predicts the nonlinear behavior in the frequency-magnitude distribution, the decrease of the b-slope associated with the asperity-like events, the magnitude range of major earthquakes, and their recurrence times.

Key words: Discrete source model, characteristic earthquakes, nonlinear frequency-magnitude distribution, seismic cycle.

Introduction

The stress inhomogeneities play an essential role in the dynamics of the geophysical processes over a wide lengthscale domain, from tectonic plates to microfractures. Seismic activity appears to be clustered in different blocks, somewhat independent from each other, which in turn also have an inhomogeneous structure with high and low strength regions, such as barriers or asperities. The hypothesis of a self-similar distribution of inhomogeneities over the fault plane has been widely adopted by seismologists to model the earthquake generation. Although it holds well for the worldwide seismicity, numerous observations emphasized significant deviations from it when restrained to regional scale (e.g., Scholz and Aviles, 1986; Main and Burton, 1986; Trifu and Radulian, 1989). Thus, a magnitude gap is noticed in the frequency-magnitude distribution between moderate and large earthquakes, followed by an enhancement at largest magnitudes. Meanwhile, Aki (1987) pointed out the constancy of the seismic moment below a certain low magnitude which might be related to the presence of an elementary seismic source of a typical size.

---

[1] Center of Earth Physics, P.O. Box MG-2, Bucharest, Romania.
[2] Institute for Physics and Nuclear Engineering, P.O. Box MG-6, Bucharest, Romania.

The nonself-similar behavior at large magnitudes may be modeled by a percolation process. In this view, LOMNITZ-ADLER (1988) developed an asperity model which includes three characteristic earthquakes: small events rupturing a single elementary surface, intermediate events rupturing an asperity coupled with a finite cluster of weak shear stress surface, and major events rupturing an asperity coupled with a percolated cluster. Correspondingly, he obtained three regions in the frequency-magnitude distribution: two peaks at low and high magnitudes, respectively, and a linear decrease between them. This model predicts rather well the seismicity features for long time intervals as compared with the seismic cycle duration, namely the seismic quiescence after a major event, major earthquake recurrence times, magnitude gap, etc. However, it is less appropriate for smaller time intervals and smaller events. Since the total asperity area occupies no more than 0.5% of the seismic active surface able to generate a major event, an anomalous high number of small earthquakes is obtained, while the linear intermediate part of the frequency-magnitude distribution is due only to events occurring close to percolation.

TRIFU and RADULIAN (1991) proposed another percolation model, based on seismotectonic data of the Vrancea (Romania) intermediate depth region. It is an attempt to cover a broad range of magnitudes, from microearthquakes to the largest earthquakes, and therefore different scalelengths in time. This model overcomes the shortcomings in the range of small earthquakes which Lomnitz-Adler's model cannot address. The observation data allowed a clear delineation of two kinds of seismicity, background crack-like earthquakes and asperity-like events, and consequently the emphasis of the asperity threshold, and a better setting of the ratio between background and asperity-like events.

The present study brings quantitative arguments in favor of the latter model by the numerical simulation of the seismic process within a seismic cycle. The algorithm was made by CĂRBUNAR (1991) and includes a series of constraints which determine a significant departure from pure percolation.

Model Parameters

The analysis of both a 9-year microearthquake catalog ($M_L \geq 2.6$) and a 50-year earthquake catalog ($M_S \geq 4$) for the Vrancea intermediate depth seismic region ($h > 60$ km), led TRIFU and RADULIAN (1991) to the conclusion that the frequency-magnitude distribution has undoubtedly a nonlinear behavior. Typical distributions in this region are shown in Figures 1 and 2. These shapes are better indicated when restraining to a particular active zone (a fault segment able to generate a major earthquake) and tend to disappear when several zones are gathered up, due to a summation effect.

The seismic process is imagined as follows: the active zone develops a discrete structure (elementary surfaces) as a characteristic feature of the fragmentation of

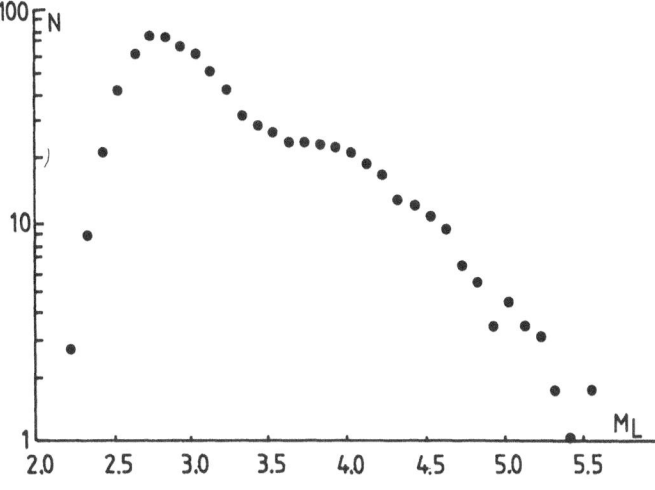

Figure 1
Noncumulative frequency-magnitude distribution of microearthquakes occurred between Oct. 1980 and
Aug. 1986 in the Vrancea region, $h > 115$ km

the lithosphere at the earthquake scale. A distribution of asperity-cells (high strength elementary surfaces) is assumed, and an asperity represents a cluster of asperity-cells. The weaker areas will fail by a crack-like mechanism, creating this way a shear stress-free surface that will tend to surround the harder areas (asperity-cells), which will rupture by a different mechanism (asperity-like). When the shear stress-free surface reaches a critical value, a percolated cluster suddenly appears and the inception of a major earthquake is likely to occur from now on.

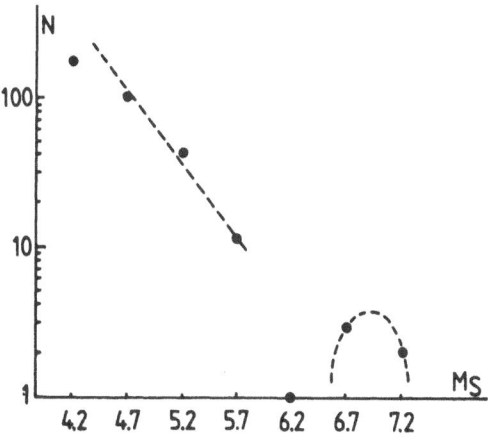

Figure 2
Noncumulative frequency-magnitude distribution of Vrancea intermediate depth earthquakes ($h > 60$ km) occurred between 1936 and 1986.

Percolation produces a magnitude gap in the frequency-magnitude distribution between moderate and large events (Figure 2). The intermediate bump at magnitude $M_L \sim 4.0$ (Figure 1) is a crucial effect to be considered. It is associated with the threshold of the asperity-like events, as a consequence of the jump in the slip surface from the crack-like event, which affects only an elementary surface, to the asperity-like event, which requires a minimum low shear stress surface around an asperity-cell. This is a remarkable effect of the Vrancea intermediate depth earthquakes, which is also emphasized in their displacement spectra showing two corner frequencies (TRIFU, 1987), and their waveform complexities (TRIFU and RADULIAN, 1989).

In the following a 2D numerical simulation is performed for the seismic process on an active zone situated in the Vrancea lower lithosphere, roughly between 120 and 160 km depth, which hosted the major earthquakes of 1940, November 10 and 1986, August 30. We choose a square elementary-cell of linear dimension $\Delta L = 0.65$ km, whose surface equals the elementary surface of the active zone estimated by TRIFU and RADULIAN (1989) from seismic spectra. The active zone area comprises 80×70 elementary surfaces (52×45 km^2). Among them a number of some 1300 stronger elementary surfaces is randomly distributed initially. They represent asperity-cells, whose random generation leads to a certain initial distribution of asperities, consisting of single or grouped cells, the potential seeds of future asperity-like earthquakes. Due to aggregation, the number of asperity nuclei is certainly less than the number of asperity-cells. The multitude of asperity-cells is required by the considered cycle duration, ~ 40 years, and the observed occurrence rate, 15–20 events/year, for $M_L \geq M_L^a = 3.9$, where M_L^a is the asperity threshold magnitude.

It is assumed that the interaction between neighbouring asperity-cells leads to an increase of their strength. The higher the asperity strength is, the larger the surrounding stress-free surface must be, in order to increase the effective stress on the asperity up to its failure. We attribute a radius of action to each N-cell asperity group as a power-law dependence

$$R(N) = \alpha 10^{\beta N} \Delta L \qquad (1)$$

where R is the radius of a group of N linked cells, and α and β are constants. We consider that only the free surface occurrence within this radius may lead to the increase of the effective stress on the asperity. The constants $\alpha = 1.62$ and $\beta = 0.093$ were determined from the limit conditions: $R(1) = R_0$, the radius associated with the slip surface of the asperity-like earthquake of minimum magnitude (one asperity-cell), and $R(N_{max}) = R_{max}$, the radius associated with the entire active zone. By analysing spectra of local earthquakes, TRIFU (1987) obtained

$$R_a^2/R_0^2 = 1/4 \qquad (2)$$

where R_a is the asperity-cell radius, similar to the equivalent radius of the elementary surface $(R_a \simeq R_e)$, so that $R_0 = 2R_e$. For the rectangular grid we took $R_0 = 2\,\Delta L$. It was considered that $N_{max} = 14$ defines the maximum size of a major asperity which may generate a major earthquake, and $R(14) = 40\,\Delta L$. Since a group of asperity-cells is unlikely to be compact, such a major asperity has a linear dimension of the order of a few km, in agreement with observations in the region (a radius of 7.5 km was found for the asperity of the 1986, August 30 event by TRIFU, 1990).

The background events are equally likely to occur throughout the active zone. The occurrence of an asperity-like event is conditioned by the presence of an asperity and a surrounding weak area. When, inside the area limited by the radius of action (1), sufficient shear stress-free surface develops, linked to the asperity, the asperity fails and slip spreads over the corresponding free surface.

There is a minimum magnitude earthquake able to generate an elementary free-surface (and all smaller events are accompanied by a complete locking of their slip areas). According to TRIFU and RADULIAN (1991), if the static stress drop is equal in the case of both crack- and asperity-like earthquakes, then

$$M_L = M_L^e + 3/2[\log(S/S_e)]/c \tag{3}$$

where M_L is the local magnitude of an asperity-like earthquake with a slip surface S, moment calibrated for Vrancea intermediate depth earthquakes, M_L^e is the minimum magnitude of the background events (crack-like) having a slip surface S_e (elementary surface), and c is the slope of the log moment-magnitude calibration ($c = 1.0$). Using (2) and $M_L = M_L^a = 3.9$ it results that $M_L^e = 3.0$. The same authors found the background seismicity rate to be quasi-stationary, ~ 50 events/year, either before or after a major event, excepting the aftershock interval. The inference of the slip surface S by the numerical model is described in the next section.

Computation Algorithm

The model prescribes an initial distribution of asperities and a constant occurrence rate of the background events. They are both randomly generated on the active zone by using the same random function.

During the cycle, the grid cells may have three levels of strength: high, low, and zero. We consider that the seismic cycle begins with the locking of the active zone following a major earthquake. Thus, initially, the active zone contains only unruptured cells of two types: stronger (asperity-cells) and weaker, that we shall further designate as black and white cells, respectively. Any white cell may fail during the cycle by a crack-like mechanism which releases the shear stress on its surface. This way, a third type of cell appears, denoted as a gray cell. The development of the gray cells, or in other words of the shear stress-free surface, is given by the background seismicity. An asperity-like event occurs by the rupture of an asperity, represented by a single or a group of adjacent black cells (black cluster), its rupture being possible only if it is surrounded by a certain shear stress-free surface or a number of gray cells.

The initial configuration of black cells is done by a random function, and consequently depends on the chosen seed. The introduction of each black cell in the grid implies two stages:

> (i) communication to its neighbours, so that every cell will know at every trial the structure of its neighbours; since a rectangular, isotropic grid is considered, we assume 8 directions of interest (Figure 3);
>
> (ii) its position in a cluster of connected black cells, with three possibilities: (a) the black cell is isolated and becomes the leader of a black cluster; (b) it is adjacent to a black cluster; (c) it is adjacent to several distinct black clusters which it connects in a single cluster (Figure 4).

For a cycle of about 40 years and a rate of asperity-like earthquakes of 15–20 events/year, a number of about 750 black clusters is initially required. Independently of the random function seed, the problem geometry imposes a saturation of the black cluster number when increasing trials, due to aggregation. In our case this limit is reached for about 1000 trials and is somewhat more than 400 clusters (composed on average of a total of about 900 black cells). To increase the number of initial black clusters, the random trials are continued, but only isolated, single-cell black clusters are considered (possibility (a) of the previous item (ii)). This procedure finishes when no more isolated single-cell black clusters can be injected into the grid, and leads to a total number of about 850 black clusters. This value is statistically independent of the chosen random function seed and covers not only the number of asperity-like earthquakes that the Vrancea region requires, but also some possible simultaneous failures (multiple events), and a number of asperities that possibly remain unbroken after the major earthquake. This impor-

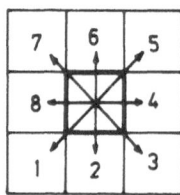

Figure 3
Elementary cell and its neighbour directions for an isotropic grid.

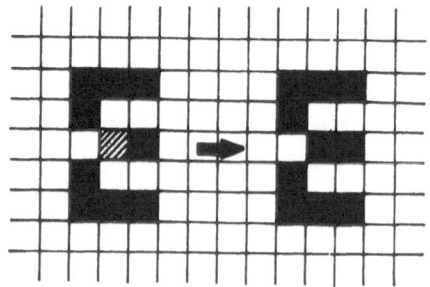

Figure 4
Linkage of three distinct black clusters by an adjacent black cell.

tant initial condition of the algorithm assures a high degree of homogeneity in the asperity distribution on the grid.

After the initial configuration is set up, the algorithm establishes the rules the cellular automaton evolves with time. The background seismicity randomly transforms white cells into gray cells, at a constant rate (50 cells/year). All white cells are included in a single white cluster, which continuously decreases with time. Simultaneously, gray cells gradually make up gray clusters. The introduction of every gray cell into the grid follows the same two stages that have been described for the generation of initial black clusters.

An information set is attached to each cell, specifying:

(i) the type of neighbours;

(ii) the index of the cluster it belongs to, for a black or gray cell;

(iii) the impending cell in the cluster.

There is a permanent account of each cluster (of every colour), containing the number of cells, the leader, and a strength index for black clusters.

We assume that the strength of an initial black cluster exponentially increases with its size

$$P(N) = \gamma 10^{\varepsilon N} \tag{4}$$

where $P(N)$ is the strength index of an N-cell black cluster, and γ and ε are determined from the limit conditions as will be further discussed. The gray surface development within a black cluster radius of action leads to the increase of the effective stress on the black cluster, which is equivalent to the decrease of its strength. The weakening algorithm considers that the occurrence of a gray cell will diminish the stength of a black cluster by an amount of ΔP only if it is linked to it by itself or through other gray cells. The corresponding adjacent gray cells to the black cluster are denoted as contact points. ΔP depends on the distance x to every contact point. Thus, the number of contact points gives a measure of the angle the gray cell sees the black cluster which it weakens, analogous to the solid angle in which it would radiate energy (Figure 5). A linear decrease with distance in 4 levels of $2 \Delta L$ each is assigned to the weakening steps, from $\Delta P = 1$ when $6 \Delta L < x \le 8 \Delta L$ to $\Delta P = 4$ when $x \le 2 \Delta L$, while $\Delta P = 0$ for $x > 8 \Delta L$. The numerical tests show that the change of this dependence has little effect on the main trends of the generation process.

Coefficients $\gamma = 8.75$ and $\varepsilon = 0.316$ in (4) were determined such as a single-cell black cluster to be most likely broken by a 3-cell gray cluster as required by (2), corresponding to an event of magnitude $M_L = 3.9$, and a 14-cell black cluster to give the strongest major event. Although much less likely, the algorithm does not exclude the situations in which a single-cell black cluster fails when only a 2-cell gray cluster is linked to it, if both gray cells are contact points, or when a large gray cluster, already made up, is suddenly connected.

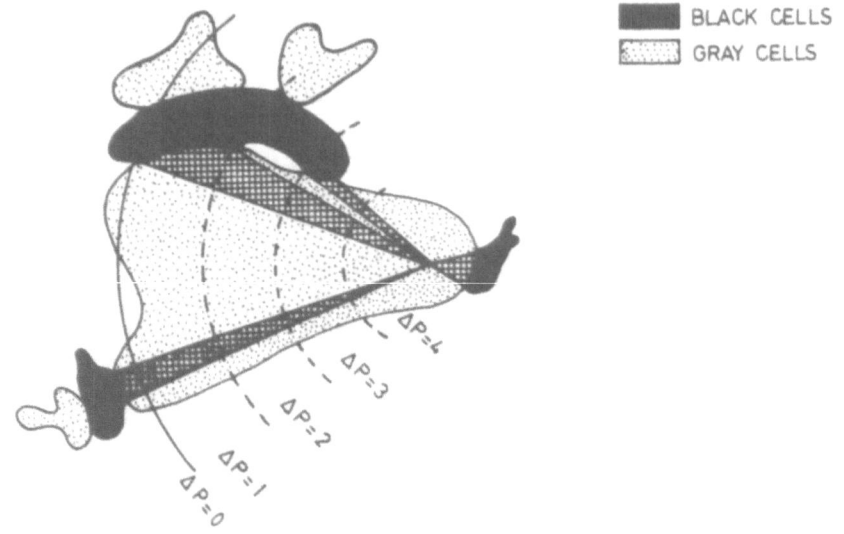

Figure 5

Black cluster strength index may diminish by the occurrence of a linked gray cell; the decrease depends on both the distance to the black cluster (in 4 decremental steps) and the angle the gray cell sees the contact points (the double hatched area).

When $P \leq 0$ an asperity-like event is triggered. Its corresponding magnitude is determined from equation (3). The surface S is computed by summing the surface of the black cluster and the surface of the linked gray cells situated within the asperity action radius, $R(N)$, measured from the mass center of the N-cell black cluster. A gray cell may simultaneously trigger the rupture of several black clusters and they will be considered as a single earthquake (multiple event). Evidently, its magnitude is obtained by accounting for the total implied surface carefully avoiding the overlapping in the computation of the gray area (Figure 6). If no locking is assumed for the asperity-like events, black cells would turn into gray cells and will consequently provoke an extremely high expansion of the gray clusters, generating the major earthquake much too early. To prevent this, a partial recovery is assumed by turning, for simplicity, the ruptured black cells into white cells, while gray cells remain unchanged during the cycle.

To identify the occurrence of the major event, two efficient criteria were statistically found: the total number of background events must overpass a critical value expressing the percolation threshold, and the magnitude of the major event must have a large enough value (say $M_L = 6.5$). Since the major earthquake has a domino effect, two features are tested when the above criteria are fulfilled. First, the temporal vicinity is analyzed previous to the triggering in order to find the occurrence of a possible event of large magnitude that will be considered as a first shock (subevent) of the major event. Second, the strength indexes of all black

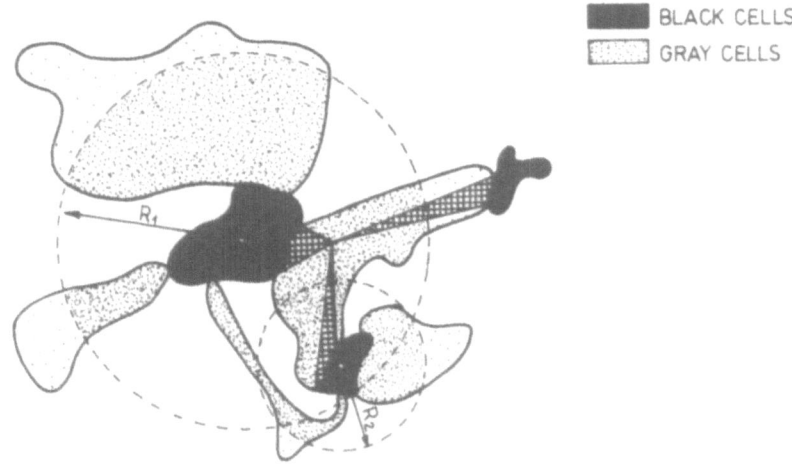

BLACK CELLS
GRAY CELLS

Figure 6
A possible configuration simulating a multiple event; the presence of the gray cell leads to the simultaneous failure of two black clusters (asperities); the slip area lies within their corresponding radii of action, $R(N_1)$ and $R(N_2)$, avoiding the overlapping.

clusters are analyzed at the time of the major earthquake. Every such cluster is appreciated to be triggered by the major event if its present strength is less that 25% of the initial value, and is consequently included in the computation of the major magnitude. If physical conditions are available for incorporation into the algorithm, it will be possible to delineate the characteristics of different failure patterns (foreshock, doublet, multiple event).

Results

To illustrate the proposed numerical procedure, Figure 7 presents the manner in which the clusters of black and gray cells evolve on the active zone from the initial locked fault to percolation. The input asperity (black clusters) distribution of this example is given in Table 1 together with the corresponding strength indexes; the number of clusters is 413. A uniform single-cell black cluster distribution is superimposed on the original 80×70 grid, providing another 448 clusters, as described in the previous section. The total number of asperities (861) is fully determined by the problem geometry. The resulting distribution is shown in Figure 7a.

Figure 7 comparatively shows the time evolution of black clusters (frames a, c, e, g) and gray clusters (frames b, d, f, h) on the grid at different stages of the cycle. The grid is randomly filled with gray cells at a rate of 50 cells/year. When the gray area increases, the probability of a black cluster failure increases too, and the number of black clusters continuously decreases. The size of the asperity-like events

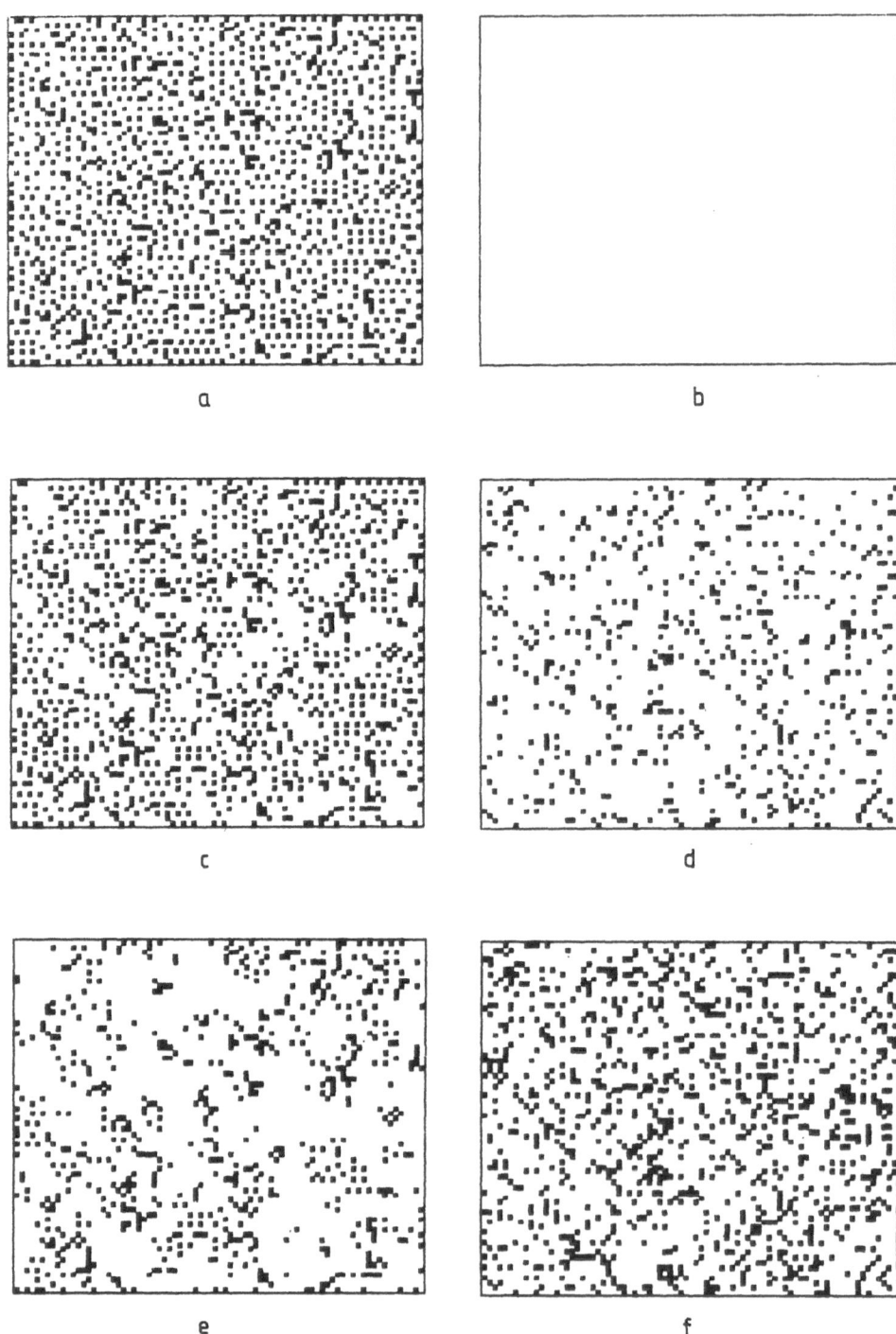

<table>
<tr><td></td><td></td></tr>
</table>

a   b

c   d

e   f

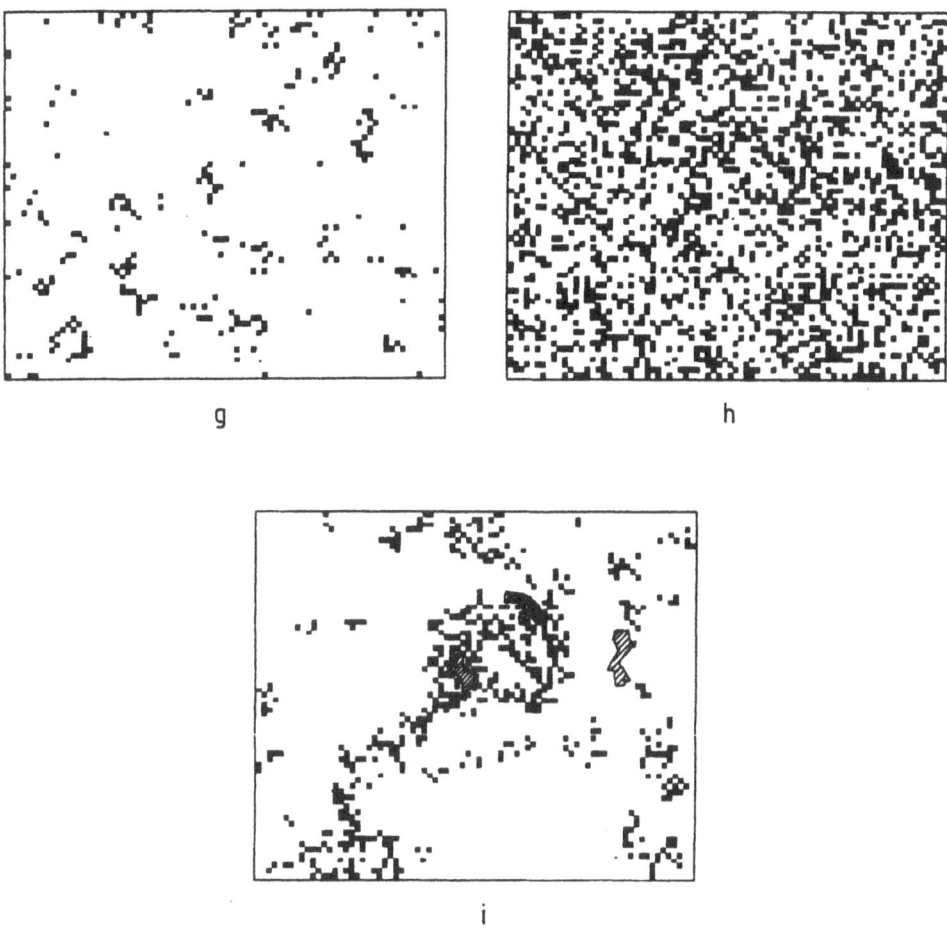

Figure 7

Evolution of black and gray clusters (left and right frames, respectively) during the cycle: black clusters represent asperities, and gray clusters represent shear stress-free surfaces; (a, b) initial configuration; (c–f) configurations at one third and two thirds of the cycle; (g, h) final configuration at triggering; (i) major earthquake: associated major black clusters represented by hatched surfaces, surrounded by gray cells defining the slip area (the right sided 14-cell asperity does not fail).

plotted as a function of time (Figure 8) indicates an increase in the average earthquake magnitude as the percolation approaches. Although such an effect was pointed out in some seismic regions (e.g., MOGI, 1979), it is not a striking feature in Vrancea.

The major event is triggered at about 35 years from the cycle inception, at a free surface concentration of about 32%, by the failure of both a 10-cell and an 11-cell asperities. They are drawn in Figure 7i, together with the slip surface (gray cells) involved in the major earthquake. It merits mentioning in this case that the largest

Table 1

Initial random distribution of asperities and their corresponding strength indexes

Number of cells	1	2	3	4	5	6	7	8	9	10	11	12	13	14
Number of clusters	216	90	42	22	15	10	4	4	4	2	3	0	0	1
Weights		12	16	23	31	42	58	80	110	150	206	283	388	532	730

asperity (14-cell black cluster) does not break during the cycle. Randomly, a large enough gray surface surrounds rather fast the somewhat smaller asperities, and creates the necessary conditions for the inception of the major event. This also explains the relatively low percentage of the free surface as compared with 44%, the critical value for a pure percolation without constraints (STAUFFER, 1979). The performed tests revealed that the average percentage for triggering is around 36% in our model.

Figure 9 presents the time variation of the occurrence rate for the asperity-like events. There is a large time interval of quasi-constant rate (~ 20 events/year). Due to initial complete locking of the fault there is a time lapse at the beginning of the cycle which is necessary to develop enough free surface to attain the stationary regime. A tendency of the asperities to exhaust arises when approaching the major

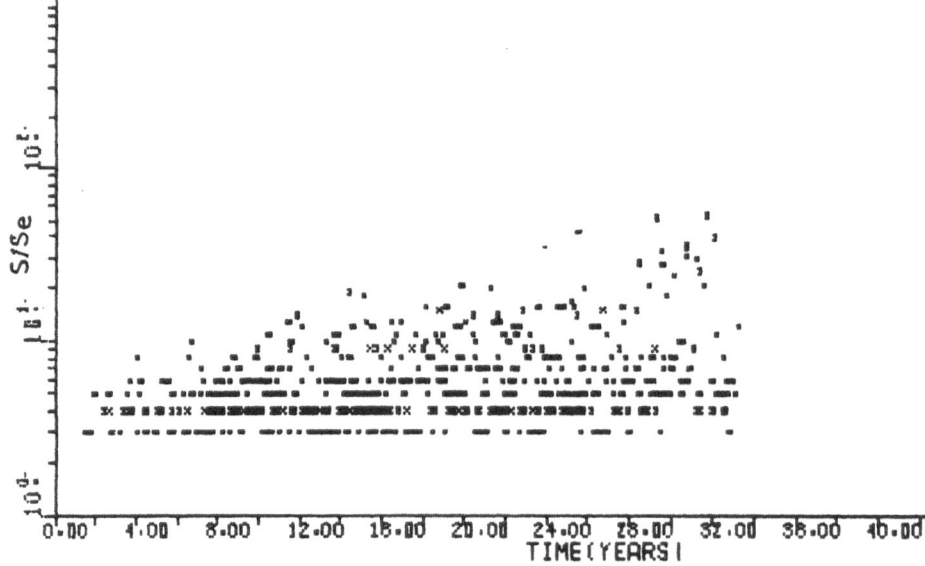

Figure 8

Earthquake size dependence with time within a cycle; every point represents an asperity-like event with a source area S; (S_e is the elementary cell area).

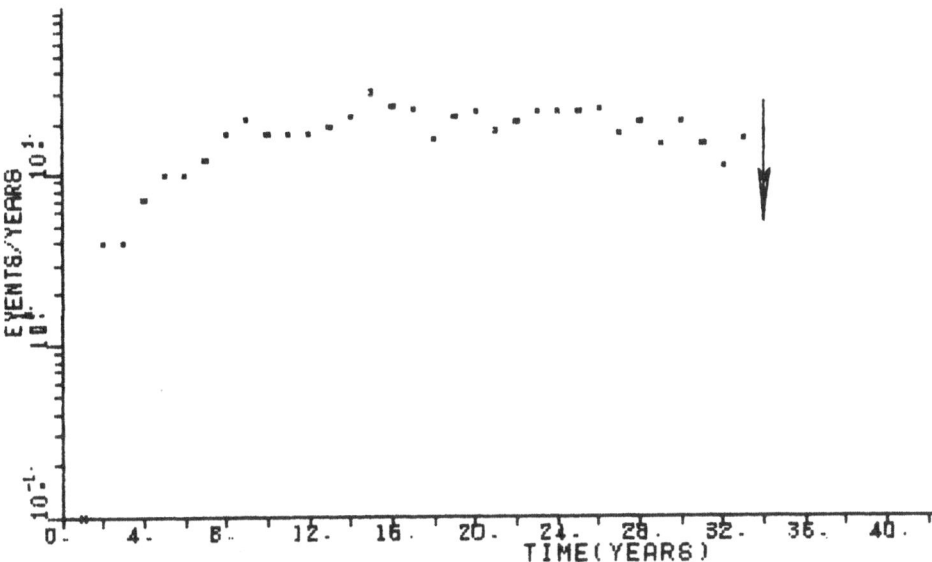

Figure 9

Occurrence rate of the asperity-like earthquakes vs. time; the arrow points to the major event.

event, implying a slight decrease. The later the major earthquake occurs, the more prominent this decay is.

The frequency-magnitude distribution of the asperity events for the above simulation is presented in Figure 10. $N(M)$ is a noncumulative value averaged on an interval of 0.2 magnitude units centered on M. This smoothing was introduced due to the presence of a numerical modulation in the distribution caused by the discrete jumps in source areas. A magnitude gap is noticed between moderate and large magnitudes. This feature fits very well the observation data at the scale of the seismic cycle (Figure 2). Moreover, since the small-scale parameters were chosen from observations (background seismicity rate, asperity threshold, elementary cell area and active zone surface) the modeling is implicitly suited at small magnitudes.

Discussion

The present study proposes a numerical algorithm for simulating the seismic process based on a 2D geometrical model. The coupling between the tectonic stress and the fault is assumed to be determined by the asperity distribution and the generation of shear stress-free surface, which are essentially governing the seismic activity. In our view, every seismic region may be divided into several active zones, each of them being characterized by a background activity, a magnitude gap in the frequency-magnitude distribution and a major earthquake size. Besides the above

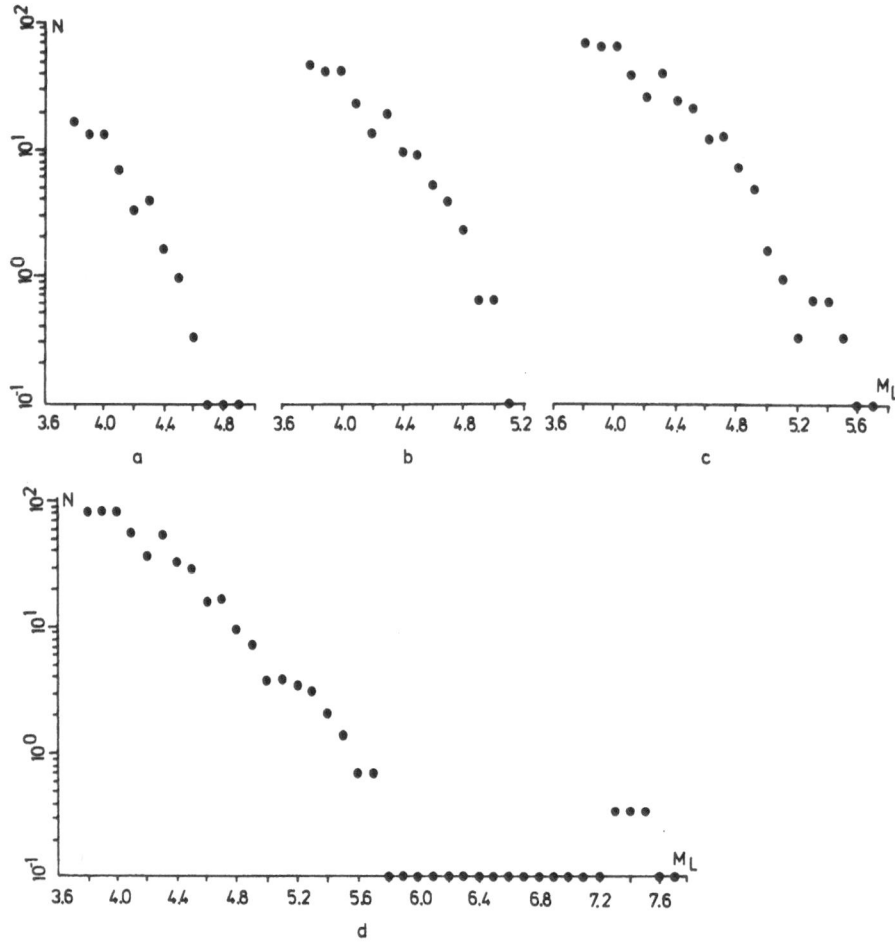

Figure 10

Simulated noncumulative frequency-magnitude distribution for different time intervals τ measured from the beginning of the cycle; (a) $\tau = T/4$; (b) $\tau = T/2$; (c) $\tau = 3T/4$; (d) $\tau = T$; T is the cycle duration; $N(M_L)$ is the average number of events on 0.2 magnitude intervals centered on M_L.

features, observed in several regions (e.g., SCHOLZ and AVILES, 1986), there are a few others which are only emphasized in the Vrancea region (TRIFU and RADU-LIAN, 1991): (i) a steady-constant level of the background seismic activity (crack-like events), and (ii) a magnitude threshold defining the inception of the asperity-like events.

The important role of stress inhomogeneities (barriers and asperities) is clearly apparent in the Vrancea region data, even at small magnitudes. Similar to crustal faults, these inhomogeneities determine an effective contact area, which is consider-ably larger in case of intermediate depth events due to the higher lithostatic stresses.

Consequently, this requires a major difference in our model with respect to Lomnitz-Adler's: a much larger percentage of asperity area contributes to the strength coupling across the fault plane (24% in comparison with 0.5%).

The algorithm assumes a discrete structure of the seismic process, as follows: (i) an elementary area for both shear stress-free surface and asperities; (ii) an asperity is comprised of a discrete number of linked asperity-cells which interact with each other. The existence of a minimum source area is implicitly involved. It might be related to a critical instability length (DIETERICH, 1986), but we prefer the assumption of a specific mechanism, such as a phase transition, or an alternative mechanism, far more likely to occur at intermediate depths. A metastable state of the lithospheric material could generate a steady-state seismic radiation at small magnitudes. Since the confining stress is high enough, a viscous layer which cannot further escape is suddenly likely to occur during the failure. On the one hand, it would explain the crack-like mechanism of the background earthquakes, and on the other hand, by its low frictional stress, the build-up of the shear stress-free surface on the fault. The presence of an asperity allows the stress accumulation within a radius of action, depending on its strength. The asperity fails when the free surface percentage inside the action area exceeds a certain critical threshold. Since the asperities act as inhibition nuclei in the spreading of the free surface, our simulation includes several constraints and is not similar to a pure percolation process (STAUFFER, 1979).

Several effects are considered to be negligible in the present modeling: (1) the overlapping, which means that no surface area slips twice; (2) the strength recovery at the scale of background earthquakes, which means that the free surface remains in this state throughout the cycle; (3) the triggering due to the stress concentration at source edges (for small and moderate earthquakes); on the contrary, this effect is fully considered in case of the major event, when the entire active zone is in an unstable state (the domino effect). Note that a change of these conditions can be further incorporated. To study the foreshock and aftershock sequences, a careful analysis of the weakening and recovery rates of the stress balance must be performed.

In conclusion, this study offers a numerical algorithm able to predict a nonlinear frequency-magnitude distribution, a gradual decrease during the cycle of the b-slope associated with the asperity-like events, the magnitude range of major earthquakes, and their recurrence times. Although the modeling includes a few physical hypotheses, it has a predominant geometrical character. It emphasizes the significant role of geometry in the mechanism of earthquake failure. Furthermore, we consider it necessary to include several other physical constraints concerning the interaction between adjacent active zones and asperities on the fault, the coupling of the shear stress-free surfaces, the stress recovery mechanism following the catastropic failure, etc. A deeper physical insight is meanwhile required to explain the discreteness of the lithosphere at the earthquake scale, and the mechanism of the background events.

REFERENCES

AKI, K. (1987), *Magnitude-frequency Relation for Small Earthquakes: A Clue to the Origin of f_{max} of Large Earthquakes*, J. Geophys. Res. *92*, 1349–1355.

CĂRBUNAR, F. O. (1991), *An Algorithm for a Percolation Process with Constraints*, in preparation.

DIETERICH, J. H. *A model for the nucleation of earthquake slip*, In *Earthquake Source Mechanics*, Maurice Ewing Ser., vol. 37 (eds. S. Das, J. Boatwright, and C. H. Scholz) (AGU, Washington, D.C. 1986) pp. 37–49.

LOMNITZ-ADLER, J. (1988), *The Theoretical Seismicity of Asperity Models: An Application to the Coast of Oaxaca*, Geophys. J. *95*, 491–501.

MAIN, I. G., and BURTON, P.W. (1986), *Long-term Earthquake Recurrence Constrained by Tectonic Seismic Moment Release Rates*, Bull. Seismol. Soc. Am. *76*, 297–304.

MOGI, K. (1979), *Two Kinds of Seismic Gaps*, Pure and Appl. Geophys. *117*, 1172–1186.

SCHOLZ, C. H., and AVILES, C. A. (1986), *The fractal geometry of faults and faulting*, In *Earthquake Source Mechanics*, Maurice Ewing Ser., vol. 37 (eds. S. Das, J. Boatwright, and C. H. Scholz) (AGU, Washington, D.C. 1986) pp. 147–155.

STAUFFER, D. (1979), *Scaling Theory of Percolating Clusters*, Phys. Rep. *54*, 1–74.

TRIFU, C-I. (1987), *Depth Distribution of Local Stress Inhomogeneities in Vrancea Region, Romania*, J. Geophys. Res. *92*, 13878–13886.

TRIFU, C-I. (1990), *Detailed Configuration of Intermediate Seismicity in Vrancea Region*, Rev. Geofis. *46*, 33–40.

TRIFU, C-I., and RADULIAN, M. (1989), *Asperity Distribution and Percolation as Fundamentals of an Earthquake Cycle*, Phys. Earth Planet. Inter. *58*, 277–288.

TRIFU, C-I., and RADULIAN, M. (1991), *Frequency-magnitude Distribution of Earthquakes in Vrancea Region: Relevance for a Discrete Model*, J. Geophys. Res. *96*, 4301–4311.

(Received March 20, 1991, revised/accepted August 14, 1991)

PAGEOPH, Vol. 136, No. 4 (1991)

0033–4553/91/040515–13$1.50 + 0.20/0

Intermagnitude Relationships and Asperity Statistics

A. A. GUSEV[1]

Abstract —Several data sources appeared recently which enable one to construct an updated version of global average intermagnitude relationships. A set of nonlinear magnitude vs M_w (or log M_o) curves is presented. Several regional scales are also included in the set. Utilization of M_w as a referential scale provides the optimal basis for extrapolation of return periods, strong motion amplitudes and source parameters from moderate to large earthquakes. Remarkable features of the constructed curves are: (1) practical coincidence of modified m_b (m_b^* or $\hat{m}_b$) scale with m^{SKM} of the Soviet Seismological Service (except the constant shift of 0.18); (2) the lack of true saturation of all scales but m_b with the possible exception of m_B for $M_w > 8.7$; (3) almost common shape of curves for all short-period magnitudes (global as well as regional); (4) lack of a systematic world-averaged difference between American and Soviet M_s.

For M_o large enough, m^{SKM} (or m_b^*) vs log M_o trend has the slope $b = 0.35$. Data are compiled to estimate short-period spectral level vs log M_o trend, and its slope seems to be near $\beta = 0.39$. These two values and some additional assumptions, rather common, lead to the following conclusions regarding properties of the earthquake source and its short-period radiation: (1) peak to rms amplitude ratio (peak factor) of a short-period record grows as $M_o^{0.11-0.13}$, hence the Gaussian noise model of the record (predicting $M_o^{0.03-0.04}$) can be definitely rejected, and some heavy-tailed peak distribution is to be assumed instead; (2) if one specifies this distribution to be the power-law (Pareto), then the estimate of the exponent α for this distribution approaches 2 for short-period teleseismic records (and this α value is the same as had been found previously for accelerogram peaks); (3) this may indicate the power law distribution of seismic force values of individual asperities; (4) there is a difference between our estimated $\beta = 0.39$ and $\beta = 1/3$ expected from the ω^{-2} model; data are not sufficient at present to show definitely that the difference is real.

Key words: Magnitude, moment magnitude, saturation of magnitude, peak factor, asperity, seismic force, Pareto distribution.

Introduction

Many practical problems of seismology require conversion between different magnitude scales. Empirical intermagnitude relationships were initially approximated as linear but detailed investigations revealed nonlinear relationships in most cases. UTSU (1982) systematically compared magnitudes based on moment magnitude $M_w = 2/3 \log M_o - 10.7$ (HANKS and KANAMORI, 1979) as the fundamental

---

[1] Institute of Volcanology, Petropavlovsk-Kamchatsky, 683006, USSR. Now at Institute of Volcanic Geology and Geochemistry, Petropavlovsk-Kamchatsky, 683006, USSR.

one. GUSEV (1983) used nonlinear $M - \log M_o$ relations to study spectral trends. Data acquired presently enable the improvement of these results. Routine M_o determinations by the Harvard group (DZIEWONSKI and WOODHOUSE, 1983) and the NEIC group (SIPKIN, 1986) provided additional data for the derivation of updated intermagnitude relations for small to moderate events. For large events, a very useful data set was compiled by PURCARU and BERCKHEMER (1982). The data set published by ABE (1981, 1984) and ABE and NOGUCHI (1983a,b) is also very important. The four main data sets will be denoted below as HAR, NEIC, PB and A. Conventional catalogues of the Soviet ESSN and American NEIC services are used as well, and some published intermagnitude relations are also included in compilation.

Throughout the paper we use the M_w (or M_o) scale as the basic one. This scale is physically transparent, and predictions of fault/source parameters and of strong motion amplitudes based on it have more chances for success. Below, we shall limit our study to shallow focus earthquakes only; moreover, condition $h < 50$ km was applied when studying M_s.

The results of our study of intermagnitude relationships are mainly of technical value. Indeed, they provide powerful tools to study radiation properties of "average" earthquakes. In the concluding part of the paper we show some implications of such an approach.

A Technique to Estimate Intermagnitude Relationships. Notation

To determine a linear intermagnitude relationship, a method of orthogonal regression is normally used. No standard nonlinear generalization of this technique was at our disposal, thus we employed the following procedure (designed to imitate the orthogonal regression locally): (1) choose the scales of the plot of two magnitudes (M_1 vs M_2) so that the data dispersion is nearly equal along both axes; (2) plot the data points; (3) rotate the axes so as to exclude a linear trend; (4) approximate the data by a smooth curve. This procedure is efficient for monotonous, slightly curved regression lines (and may fail in more general cases). It can, however, produce major errors if data for one of the two magnitudes are truncated. A typical case is the m_b (M_s) curve, which appears to be steeper than in reality at low M_s due to actual truncation of M_s data at about $M_s = 5$. We found no remedy for such cases and simply stopped our procedure when truncation seemed to be substatial.

Note that even in the case of linear regression, the orthogonal regression approach is not in fact theoretically founded, and presents only some reasonable recipe. In theory (e.g., YORK, 1966), deviations of data points from a "true" $y(x)$ line are related to "inner" errors in x and y, and not to real deviations of sufficiently accurate values from the linear relation as in our case.

Surface wave magnitude M_s (either by vertical or by horizontal instrument) will be denoted as M_s^{GR} in the case employing the Gutenberg formula, M_s^{US} in the case of the "Prague" formula with $T = 17-23$ s and corresponding amplitude (NEIC practice), and M_s^{OB} in the case of the "Prague" formula with maximum amplitude and corresponding period (ESSN practice, OB for Obninsk). Bodywave magnitude values determined by medium/long-period instruments are denoted as m_B as in PB. For magnitudes determined by short-period instruments we use m_b in the case of the Benioff instrument (NEIC), and m^{SKM} in the case of the SKM-3 instrument (ESSN); M_L is the original Richter magnitude based on the 1 s instrument, and M_{JMA} is the magnitude provided by the Japanese JMA service, based on the 5 s instrument. K^{R60} and K^{F60} are energy class scales after Rautian (RIZNICHENKO, 1960), used in many regions of the continental USSR, and after FEDOTOV (1972), used in Kamchatka. These are based on 0.7 s and 1.2 s instruments, correspondingly.

Long- and Medium-period Magnitudes

The relationship M_s^{GR} (M_w) for large magnitudes was constructed based on M_s data of A for 1916–1980 and M_w of PB. Consequently it was reduced to M_s^{US} (M_w), based on the relation $M_s^{US} = M_s^{GR} + 0.18$ (see ABE, 1981), and the analysis was further carried out for M_s^{US} only. For small to moderate magnitudes we used catalogue M_s and M_o data of HAR and HEIC. The result is shown in Figure 1.

The relationship M_s^{OB} (M_w) was constructed in a similar way for small to moderate earthquakes. For large events data are scarce. The hypothesis $M_s^{OB} \neq M_s^{US}$ was tested by using data from 1968–1983 and consequently rejected. In general, at $M_s > 5$, differences between M_s^{US} and M_s^{OB} are below 0.05, and we shall assume $M_s^{OB} = M_s^{US}$. Thus, M_s^{OB} branch is plotted in Figure 1 for $M_s < 5$ only, where differences are significant. Up to $M_s = 8.0$ our M_s^{US} vs M_o curve is naturally identical to that of EKSTRÖM and DZIEWONSKI (1988), including the useful relation

$$\log M_o = M_s^{US} + 19.24$$

for M_s^{US} range of 4–6.

The relationship m_B (M_w) for large magnitudes was established using again m_B from A and M_o from PB. For small to moderate magnitudes, m_B data from ESSN and M_o data from HAR were used. The result is seen in Figure 1; it can be tested by the substitution of the above-determined relation M_s^{US} (M_w), with the correction of 0.18, into the formula (ABE and KANAMORI, 1980)

$$m_B = 0.65 M_s^{GR} + 2.5;$$

giving similar results up to $M_w = 8.7$. However, the observed tendency to saturation at $M_w \approx 9$ does not agree with the result of such substitution.

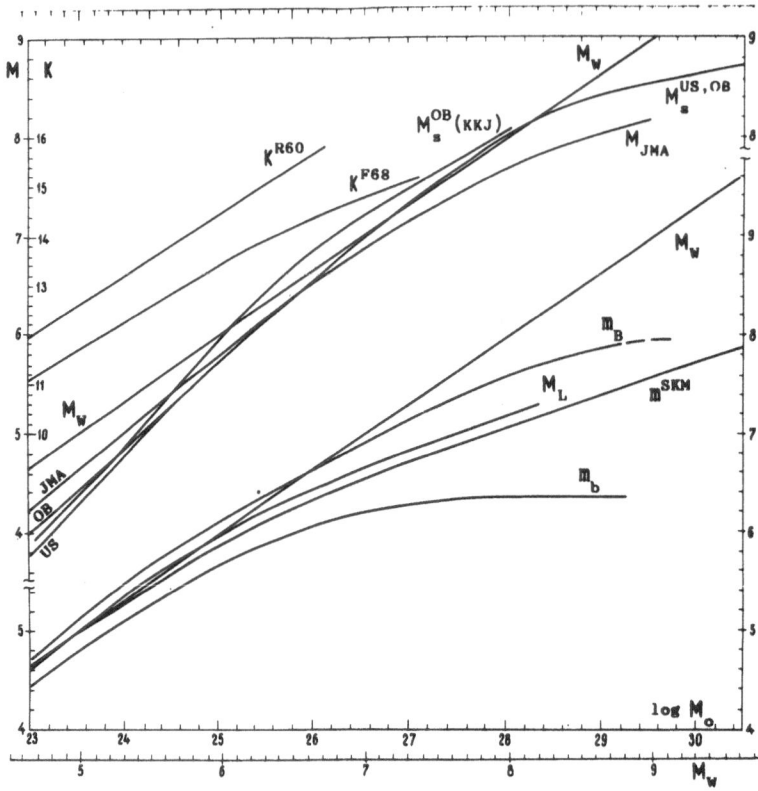

Figure 1

World average intermagnitude relationships presented with M_w (and M_o) as an argument. Regional scales K, M_L and M_{JMA} are also presented and the regional average M_s^{OB} curve for Kamchatka-Kuriles-Japan is shown too. To relate two magnitudes M_1 and M_2 one can proceed as $M_1 \rightarrow M_w \rightarrow M_2$. For notation see the text.

Short-period Magnitudes

Short-period magnitude values depend on the precise value of the instrument period. A slight difference between Benioff and SKM-3 instruments leads to a minor difference between m_b and m^{SKM}, even at $m_b < 5$ where the procedures for their determination more or less coincide. At $m_b > 5.5$, an additional requirement being used in the NEIC practice becomes essential, namely to select maximum amplitude only from several first peaks. As the true maximum normally arrives later, at large M_w, m_b lacks clear meaning and saturates at the level of about 6.4. KOYAMA and ZHENG (1985) and also HOUSTON and KANAMORI (1986) redetermined m_b values for many large earthquakes using true maxima, and they denote the corrected m_b as m_b^* or $\hat{m}_b$. At first we carried out separate regression for m^{SKM}

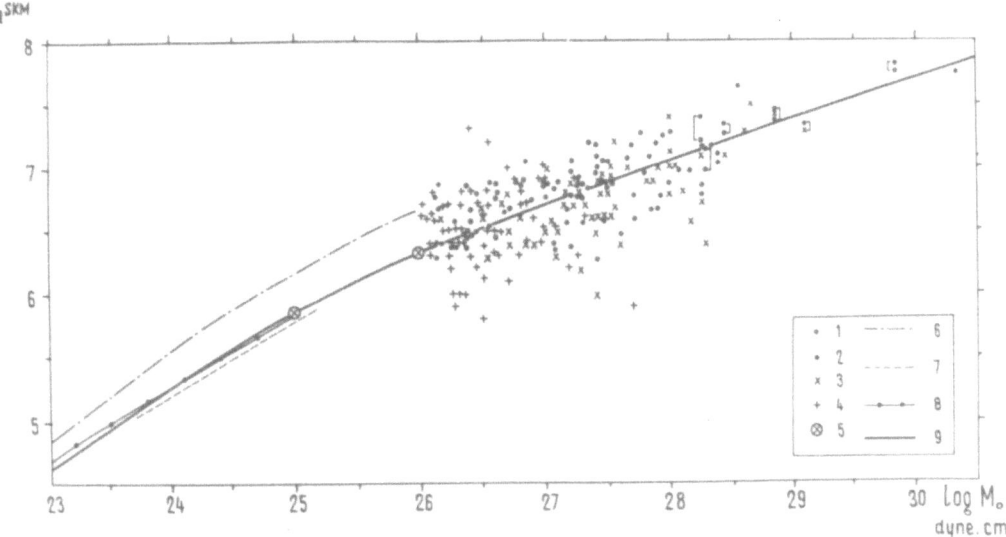

Figure 2

m^{SKM} vs M_o relationship. m_b, m_b^* and $\hat{m}_b$ data are also given, with correction of $+0.18$. 1—m_b^* after
KOYAMA and ZHENG (1985); 2—$\hat{m}_b$ after HOUSTON and KANAMORI (1986) and ZHOU and
KANAMORI (1987); 3—m^{SKM}, mostly from PURCARU and BERCKHEMER (1981); 4—m^{SKM} from
ESSN catalogue and M_o from GIARDINI et al. (1985); 5—data centroids for m_b at given M_o from HAR;
6—"continental" trend of JOHNSON (1989); 7—"plate margin" trend of NUTTLI (1985); 8—trend for
Kamchatka data partly using M_o from direct body waves; 9—accepted trend. Different estimates for the
same event are indicated by a connection line.

and m_b^*, but discovered that regression curves practically coincide except for the
constant difference

$$m_b^* = m^{SKM} - 0.18.$$

Thus, the plot in Figure 2 is given for m^{SKM} (M_o) or m_b^* $(M_o) + 0.18$. In the
$M_w = 6.6–9.5$ range it is well approximated by the straight line:

$$m^{SKM} = 0.525 M_w + 2.86 = 0.35 \log M_o - 2.75.$$

Earlier values for the slope b of this line come from KOYAMA and SHIMADA (1985),
$b = 0.40$, and from HOUSTON and KANAMORI (1986), $b = 0.37$. m^{SKM} and m_b
curves are also plotted on Figure 1, the latter being based in part on BOORE (1986).

Regional Scales

Regional scales M_L, K^{R60}, K^{F68}, M_{JMA} are used in the regions of detailed
seismological research. Their relationships versus M_w were mainly compiled. Note
that the K scale, in its widely used versions, is a magnitude-type scale based on peak

amplitudes, and its relation to seismic energy is indirect (of regression type). In Figure 1 we present the M_L (M_w) curve based mainly on HANKS and BOORE (1984), the M_{JMA} (M_w) curve based on M_{JMA} (M_w) and M_{JMA} (M_s) plots of UTSU (1982), and the K^{R60} (M_w) curve according to RAUTIAN *et al.* (1981). As for K^{F68} vs M_w, this relationship was built up by the author. To obtain M_o values for this derivation we used HAR data, M_s data converted to M_o according to the world average relation, and M_o values determined by E. M. GUSEVA from the local records of direct body waves (observed on steep rays only, with clear unipolar body wave pulse). K^{F68} values from the catalogue of the Kamchatka regional seismic net were used at $K^{F68} \leq 13$. At larger K^{F68} values, standard regional instruments are largely off scale, and K^{F68} values determined routinely were demonstrated by V. M. ZOBIN to be inaccurate. Thus, instead of K^{F68} data, at $K^{F68} > 13$ we used indirectly estimated K^{F68} values based on coda waves. The technique of such estimation was earlier developed by LEMZIKOV and GUSEV (1989) and is based on the coda wave level of a short-period instrument. To estimate the coda wave class K_c value, the following empirical formula was used

$$K_c = 1.6 \log A_c(100 \text{ s}) + \text{const},$$

where A_c (100 s) is coda amplitude (reduced or measured) at 100 s lapse time. LEMZIKOV and GUSEV (1989) have shown that $K_c = K^{F68}$, with an accuracy of about 0.4.

Intermagnitude Relations for the Kamchatka Region

The above-presented relations are world-averaged, but there are more or less pronounced regional deviations from them. In the following we look for regional relationships for Kamchatka; earlier results were obtained by FEDOTOV (1972) and VIKULIN (1983).

Relationships M_s^{OB} (M_w) and M_s^{US} (M_w) were obtained for Kurile-Kamchatka (from data sources cited above) and for Japan (same sources and UTSU, 1982). M_s^{US} deviations from the average world trend are up to 0.1, while for M_s^{OB} they reach 0.35. In both cases, differences between the two regions appear to be negligible. The relationship M_s^{OB} (M_w) is plotted in Figure 1, marked KKJ.

Relationships of m_B, m^{SKM} and m_b vs M_w for Kamchatka either coincide with the world average (m^{SKM}, m_b) or differ by a constant: regional m_B (M_w) is 0.15 above the average curve. A linear relation is found for K^{F68} when $m^{SKM} = 4.3$–7;

$$K^{F68} = 2.00 m^{SKM} + 1.68.$$

Note that K^{F68} roughly equals $2 \log(A/T)$, so that $0.5K$ is somewhat of a local magnitude, and the above relation indicates the proportionality of short-period local and teleseismic amplitudes despite the substantially larger bandwidth of a local record.

On the Accuracy of the Proposed Relationships

The problem of accuracy of the presented set is complicated: a worldwide data set cannot be treated probabilistically as a statistical ensemble, because of pronounced regional variations. For the regional data set of Kamchatka, the standard deviations from average nonlinear trends (vs M_w) are: $\sigma(M_s^{OB}) = 0.35$, $\sigma(M_s^{US}) = 0.20$, $\sigma(m_B) = 0.25$, $\sigma(m_b) = \sigma(m^{SKM}) = 0.30$, $\sigma(K^{F68}) = 0.65$. For K^{F68} vs m^{SKM}, $\sigma = 0.55$.

Typical parameters of worldwide data sets are $\sigma = 0.25$–0.30 for M_s (M_w), and the same, $\sigma = 0.25$–0.30, for m^{SKM} (M_w); but the actual values are strongly sample-dependent. Comparing regional and worldwide estimates one can suspect that interregional contribution to the variance is low. The actual picture is considerably more complicated because the contribution of subduction zones (mainly Pacific) to any random worldwide data set is dominating, and other tectonic types simply cannot manifest themselves. The most well-known regional effect is interplate-intraplate difference of ≈ 0.25 for m^{SKM} or m_B; it can be seen in Figure 2. We may mention that we found the variations of the same magnitude when we divided the Kamchatka sample into subregions; thus the problem repeats itself on the more local level. As for the b value of the m^{SKM} vs log M_o slope, its accuracy can be described roughly by $\sigma = 0.02$.

Spectral and Amplitude Trends of Source Radiation in the Short-period Spectral Band and Asperity Statistics

We believe that the accurately determined value of the slope b of m^{SKM} vs log M_o linear trend bears valuable information regarding properties of the earthquake source. For illustrative purposes one can compare our empirical estimate $b = 0.35$ with the theoretical value $b = 0.20$ obtained by HANKS and McGUIRE (1981) for the combination of ω^{-2} spectral model and the Gaussian process model of a record. Though they strove for a trend of peak acceleration, their result holds true for teleseismic amplitudes as well, as the numbers of peaks are comparable. The difference between the empirical and the theoretical estimates is drastic and it follows that at least one of the two theoretic assumptions must be rejected.

A short-period teleseismic P-wave record, the peak of which determines m^{SKM}, can be considered as a segment of narrow-band random signal with its central frequency close to 0.7 Hz. To study the scaling of peak amplitudes one can base on rms amplitude which can be related to the Fourier spectrum of a record, and further to the source spectrum $\dot{M}(f)$. Hence we are interested in the estimates of the empirical trend of $\dot{M}(f)|_{f=0.7\,\text{Hz}} = S_{0.7}$.

Let β be the exponent in the relationship

$$\dot{M}(f) \propto M_o^\beta.$$

Data on β are scarce. GUSEV (1983) estimated the trend of a short-period spectral level from the trend of short-period body wave magnitudes; the slope of his linear relationship for the log $M_o = 26-29$ range indicates $\beta = 0.37$ at 1 Hz. This estimate cannot now be considered as reliable because the Gaussian process model was assumed during its derivation. KOYAMA and ZHENG (1985) combined peak amplitude A_{peak} and duration d to determine the $S_{0.7}$ vs M_o trend; they obtained the slope value $\beta_{\text{KZ}} = 0.50$ for log $M_o = 25-30$. They also studied the spectral level vs m_b^* relationship and found it to be linear in the wide range $m_b^* = 5.5-7.5$:

$$\log S_{0.7} = k_\beta m_b^* + \text{const}$$

with $k_\beta = 1.24$. All their results are biased since they implicitly assumed the rms amplitude to be proportional to peak one (or peak factor to be magnitude-independent). This is not actually the case, hence the direct use of numerical values of their β and k_β values is impossible. Their observation of linearity of log $S_{0.7}$ vs m_b^* trends in a rather wide magnitude range is, however, of vital importance.

One can try, however, to propose some interpretation for these data. Denote the peak factor $PF = A_{\text{peak}}/A_{\text{rms}}$, where A_{rms} is the rms amplitude, and let

$$PF \propto M_o^p.$$

Then $\beta_{\text{KZ}} = \beta + p$ (p is nonnegative, so 0.50 and 1.24 are upper bound for β and k_β). Now note that $S_{0.7} \propto A_{\text{rms}} d^{0.5}$, and let $d \propto M_o^q$, then $A_{\text{rms}} \propto M_o^{\beta - 0.5q}$. Combining this with $A_{\text{peak}} \propto M_o^b$, obtain

$$p = b - \beta + 0.5q$$

So that $\beta_{\text{KZ}} = b - 0.5q$. At $\beta_{\text{KZ}} = 0.50$ and $b = 0.35$ this gives $q = 0.30$. This is only slightly below $q = 1/3$ which is the value that could be expected, based on the similarity assumption; the difference may represent the effect of scattering (which leads to $q \approx 0$ at small enough M_o).

Spectral trend was properly estimated first by HOUSTON and KANAMORI (1986) who found $\beta = 0.45$ for 0.55 Hz based on a limited data set. ZHOU and KANAMORI (1987) revealed large data dispersion reflecting a tectonic setting of earthquakes and found $\beta = 0.52$ for 1 Hz and $\beta = 0.41$ for 0.5 Hz for subduction zone event subgroup of their data. Regrouping their data differently, one can obtain other figures; for the subgroup of plate margin earthquakes, for instance, we found that β nears 0.40 for both 1 and 0.5 Hz. Such freedom of interpretation seems to be produced by the mentioned data dispersion as well as by a relatively narrow magnitude range.

HARTZELL and HEATON (1985) studied Pasadena spectral levels and found the nonlinear trend of spectral level for the short to medium period range. For 0.7 Hz their spectral trend can be roughly described by two linear segments: one with $\beta = 0.39$ for $M_w = 7$–8.5 (53 events), and another with $\beta = 0$–0.1 for $M_w > 8.5$ (5 events). HARTZELL and HEATON (1988) try to confirm this saturation of the spectral level by the observation of a similar saturation for amplitudes; this trend is virtually equivalent to the m_B vs M_o relationship. The observed nonlinearity is equivalent to that seen in Figure 1 for m_B. For the frequency band of around 0.7 Hz, however, the analogue of m_b^* vs log M_o relation given by HARTZELL and HEATON (1988) demonstrates no visible saturation. Data of KOYAMA and ZHENG mentioned above also show no saturation. Thus, we shall not base our estimates on HARTZELL and HEATON'S (1985) data for $M_w = 8.75$–9.5 range.

In general, the accuracy of the described β estimates is low. Since we believe that our estimate $b = 0.35$ is rather accurate, we can try to determine β more accurately as $k_\beta b$ if we manage to determine k_β accurately. One can assume *a priori* that the $S_{0.7}$ vs log M_o relationship will show a larger dispersion than m_b^* vs $S_{0.7}$, since spectral shape anomalies (including those related to tectonic setting) will be at least partially excluded in the latter case. The data generally confirm this idea: for log M_1 value of KOYAMA and ZHENG (1985) ($M_1 \approx S_{0.7}$), $\sigma(\log M_1) \approx 0.25$ on their M_1 vs M_0 plot and $\sigma(\log M_1) \approx 0.20$ on their M_1 vs m_b^* plot. Hence, certain variance reduction can truly be achieved. Thus, an indirect estimate of β can be more accurate.

Therefore, we compiled relevant data to estimate k_β. Data on $S_{0.7}$ vs $\hat{m}_b$ ($\hat{m}_b \equiv m_b^*$) from HOUSTON and KANAMORI (1986) and ZHUO and KANAMORI (1987) are plotted in Figure 3. When no m_b^* was present we used $m^{SKM} - 0.18$ instead. We make mention that plate interior earthquakes, with large positive deviations of their $S_{0.7}$ values from the regression line for subduction zone or "plate margin" groups show no anomaly on $S_{0.7}$ vs $\hat{m}_b$ plot. The absolute level of $S_{0.7}$ depends on particular assumptions made during inversion for source spectrum, mainly on t^*. HOUSTON and KANAMORI (1986) and ZHUO and KANAMORI (1987) used $t^* = 0.7$ while HARTZELL and HEATON (1985) used $t^* = 1.0$. We plotted $S_{0.7}$ values of the latter authors, averaged for several M_w intervals, both in original form and reduced to $t^* = 0.7$ according to HOUSTON and KANAMORI (1986). We could not directly display the important data set of KOYAMA and ZHENG on the same plot because of calibration problems (they used roughly $t^* = 1.4$) as well because of the aforementioned bias in their estimates. The role of these data, however, is important because of their large magnitude range. We plotted a segment of a straight line through the modified HARTZELL and HEATONS's average point for $m_b^* = 6.5$ ($M_w = 7.25$). The length of this segment represents the data range of KOYAMA and ZHENG, and its slope $k_\beta = 1.12$ equals the slope based on HARTZELL and HEATON's (1985) $\beta = 0.39$ cited above. This β value will be accepted as our final estimate.

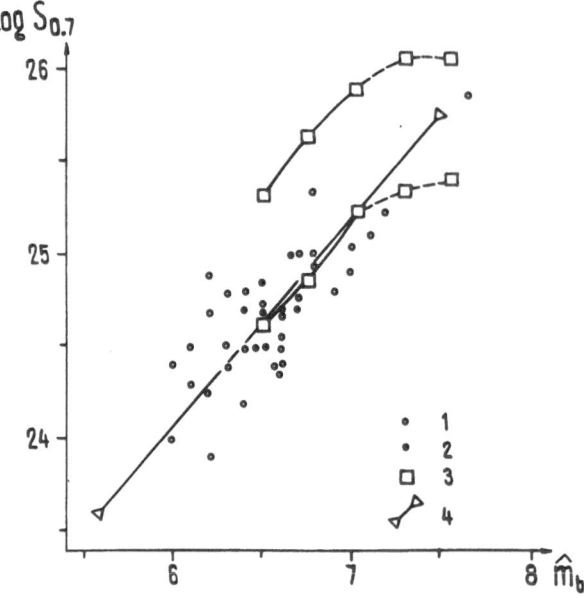

Figure 3

Source spectral level $S_{0.7}$ vs m_b relationship. $(S_{0.7} \equiv \dot{M}_o(f)|_{f = 0.7\,\text{Hz}})$. 1—data of HOUSTON and KANAMORI (1986), ZHOU and KANAMORI (1987) and HWANG and KANAMORI (1989) for plate margin earthquakes; 2—same, plate interior; 3—data centroids of $S_{0.7}$ after HARTZELL and HEATON (1988); original (the upper line) and modified by HOUSTON and KANAMORI (1986) (the lower line); 4—results of KOYAMA and ZHENG (1985), presented as a straight segment (see the text for details).

Now we can finally pass to source properties. Determining the peak factor value by the relation $p = b - \beta + 0.5q$, we obtain $p = 0.13$ for the theoretical $q = 1/3$ (or $p = 0.11$ for the empirical $q = 0.30$ based on KOYAMA and ZHENG (1985) data). Both values are fully incompatible with a Gaussian process record model which gives $p \approx 0.03$–0.04 (HANKS and MCGUIRE, 1981), and this model can be completely rejected. GUSEV (1989) assumed the power-law statistics for amplitudes of individual acceleration pulses that combine into an accelerogram, and estimated the exponent of the law to be $\alpha \approx 2$. Similarly, we assume that the teleseismic short-period record is also produced by pulses with power-law statistics. For the peak amplitude A_p, it gives

$$A_p \propto N^{1/\alpha} \propto S^{1/\alpha} \propto M_o^{2/3\alpha},$$

where N is the number of pulses. This holds, if two assumptions are true: N is proportional to the source area S, and S value is proportional to $M_o^{2/3}$. Comparing the theoretical $b = 2/3\alpha$ with the empirical $b = 0.35$, it results in $\alpha = 1.9$. From GUSEV (1989), pulse amplitude statistics of band-filtered displacement signal can represent statistics of seismic force F_o values of asperities ($F_o = \int_{S_a} \Delta\sigma\, dS$, where $\Delta\sigma$

is the local stress drop and S_a the asperity area). For an asperity population with nearly constant S_a this indicates the power law distribution of average (over S_a) $\Delta\sigma$ value, and $\alpha \approx 2$ for this law. This conclusion coincides with the analogous result of GUSEV (1989) based on near-field data.

Another interesting point is the probable difference between empirical β estimate (0.39) and $\beta = 1/3$ expected from the ω^{-2} model. One cannot be sure at present that the difference is significant, because no accurate error estimate can be ascribed to our β value. Additional evidence in support of the reality of this difference can be found in regional scaling laws. PAPAGEORGIOU and AKI'S (1985) spectral scaling, based on the Western USA data, indicates $\beta = 0.40$. SUGITO and KAMEDA (1985) analyzed the Japanese data; their results indicate $S_{0.7} \propto 10^{0.6 M_{JMA}}$ for the magnitude range $M_{JMA} = 5-8$; converting M_{JMA} to M_o by Figure 1, obtain $\beta = 0.45$. Thus we can consider the hypothesis $\beta > 1/3$ as probable. Assuming $\beta > 1/3$ one can try to explain the difference in frames of the multiasperity fault model of GUSEV (1989) if one assumes that the average asperity size $2R_a$ is slowly growing with M_o; actually $\beta = 0.39$ can mean $2R_a \propto M_o^{0.06}$.

Discussion and Conclusions

Curves of Figure 1 show several features of magnitude scales that were widely discussed in recent years: the slope value of 1 for M_s^{US}, m_B, m^{SKM}, m_b vs log M_o at low magnitudes, and saturation of m_b scale. Some new features are also emphasized:

(1) coincidence of m^{SKM} and m_b^* (modified m_b) scales up to a constant shift;
(2) lack of real saturation for all scales excepting m_b (total saturation of m_B remains under question); instead of saturation, an interval of slow increase is present at the upper end of various curves;
(3) similar shape of curves for short period scales: m^{SKM}, M_L and $0.5K^{F68}$;
(4) good agreement between M_s^{OB} and M_s^{US} (which both do not directly match either M_s^{GR} or M_{GR}). Disagreement is observed below $M_s = 5$ only, reflecting the contribution of data from small epicentral distances when inadequate visual period is used in the "Prague" formula during the M_s^{OB} determination.

To simplify applications, a tabulated version of our results is added as Table 1.

In order to compare amplitude and spectral trends for short periods we need an estimate of the spectral trend and compiled spectral data from different sources. We believe that the proposed value $\beta = 0.39$ is a reasonable starting point to deduce some conclusions regarding earthquake source. We found that the widely accepted assumption of the Gaussian noise model of short-period record is not supported by short-period teleseismic data. Some heavy-tailed distribution must be assumed to explain actual peak statistics.

Table 1

Magnitude vs seismic moment relationships: global average, for regional scales, and regional for Kamchatka-Kurile-Japan region (KKJ)

| | $\log M_o$, dyne.cm | | | | | | | |
	23	24	25	26	27	28	29	30
M_s^{GR}	3.58	4.58	5.54	6.34	7.12	7.82	8.23	8.45
M_s^{US}	3.76	4.76	5.72	6.52	7.30	8.00	8.41	8.63
M_s^{OB}	4.00	4.83	5.68	6.49	7.30	8.00	8.41	8.63
m_B	4.70	5.47	6.08	6.62	7.13	7.55	7.85	(7.98)
m^{SKM}	4.62	5.27	5.68	6.33	6.71	7.05	7.40	7.75
m_b	4.45	5.10	5.66	6.05	6.26	6.34	6.34	6.34
M_L	4.60	5.34	5.95	6.42	6.82	(7.16)		
M_{JMA}	4.22	4.99	5.77	6.49	7.12	7.64	8.04	(8.27)
K^{F68}	11.08	12.22	13.36	14.37	(15.11)	(15.80)		
M_s^{US} (KKJ)	3.73	4.68	5.65	6.47	7.25	(7.99)		
M_s^{OB} (KKJ)	3.84	4.84	5.95	6.84	7.48	(8.04)		
M_w	4.63	5.30	5.97	6.63	7.30	7.97	8.63	9.30

Acknowledgements

The author is obliged to V. N. Melnikova for technical assistance and to the anonymous reviewer for his useful criticism.

REFERENCES

ABE, K. (1981), *Magnitudes of Large Shallow Earthquakes from 1904 to 1980*, Phys. Earth Planet. Interiors *27*, 72–92.

ABE, K. (1984), *Complements to Magnitudes, 1904–1980*, Phys. Earth Planet. Interiors *34*, 17–23.

ABE, K., and KANAMORI, H. (1980), *Magnitudes of Great Shallow Earthquakes from 1953 to 1977*, Tectonophys. *62*, 191–203.

ABE, K., and NOGUCHI, S. (1983a), *Determination of Magnitude for Large Shallow Earthquakes, 1898–1917*, Phys. Earth Planet. Interiors *32*, 45–59.

ABE, K., and NOGUCHI, S. (1983b), *Revision of Magnitudes of Large Shallow Earthquakes, 1897–1912*, Phys. Earth Planet. Interiors *33*, 1–11.

BOORE, D. M. (1986), *Short-period P and S Wave Radiation from Large Earthquakes: Implications for Spectral Scaling Relations*. Bull. Seismol Soc. Am. *76*, 43–64.

DZIEWONSKI, A. M., and WOODHOUSE, J. H. (1983), *An Experiment in the Systematic Study of Global Seismicity: Centroid Moment Tensor Solutions for 201 Moderate and Large Earthquakes of 1981*, J. Geophys. Res. *88*, 3247–3271.

EKSTRÖM, G., and DZIEWONSKI, A. M. (1988), *Evidence of Bias in Estimations of Earthquake Size*, Nature *332*, 319–323.

FEDOTOV, S. A. (1972), *Energy Classification of the Kurile-Kamchatka Earthquakes and the Problem of Magnitudes*, Moscow, Nauka, 111 pp. (in Russian).

GIARDINI, D., DZIEWONSKI, A. M., and WOODHOUSE, J. H. (1985), *Centroid-moment Tensor Solutions for 113 Large Earthquakes in 1977–1980*, Phys. Earth Planet. Interiors *40*, 259–272.

GUSEV, A. A. (1983), *Descriptive Statistical Model of Earthquake Source Radiation and its Application to Short-period Strong Motion*, Geophys. J. Roy. Astr. Soc. *74*, 787–808.

GUSEV, A. A. (1989), *Multiasperity Fault Model and the Nature of Short-period Subsources*, Pure Appl. Geophys. *130*, 635–660.

HANKS, T. C., and BOORE, D. M. (1984), *Moment Magnitude Relations in Theory and Practice*, J. Geophys. Res. *89*, 6229–6236.

HANKS, T. C., and KANAMORI, H. (1979), *A Moment Magnitude Scale*, J. Geophys. Res. *84*, 2348–2350.

HANKS, T. C., and MCGUIRE (1981), *The Character of High-frequency Strong Ground Motion*, Bull. Seismol. Soc. Am. *71*, 2071–2095.

HARTZELL, S. H., and HEATON, T. H. (1985), *Teleseismic Time Functions for Large Shallow Earthquakes*, Bull. Seismol. Soc. Am. *75*, 965–1004.

HARTZELL, S. H., and HEATON, T. H. (1988), *Failure of Self-similarity for Large $(M_w > 8\frac{1}{4})$ Earthquakes*, Bull, Seismol. Soc. Am. *78*, 478–488.

HOUSTON, H., and KANAMORI, H. (1986), *Source Spectra of Great Earthquakes: Teleseismic Constraints on Rupture Process and Strong Motion*, Bull. Seismol. Soc. Am. *76*, 19–42.

HWANG, L. J., and KANAMORI, H. (1989), *Teleseismic and Strong-motion Spectra from two Earthquakes in Eastern Taiwan*, Bull. Seismol. Soc. Am. *79*, 935–944.

JOHNSON, A. C. (1989), *Moment Magnitude Estimation for Stable Continental Earthquakes*, Seismol. Res. Lett. *60*, 13.

KOYAMA, J., and SHIMADA, N. (1985), *Physical Basis of Earthquake Magnitudes: An Extreme Value of Seismic Amplitudes from Incoherent Fracture of Random Fault Patches*, Phys. Earth Planet. Interiors *40*, 301–308.

KOYAMA, J., and ZHENG, S. H. (1985), *Excitation of Short-period body Waves by Great Earthquakes*, Phys. Earth Planet. Interiors *37*, 108–123.

LEMZIKOV, V. K., and GUSEV. A. A. (1989), *Energy Classification of near Kamchatka Earthquakes by Coda Wave Level*, Volcanol. and Seismol. *4*, 83–97 (in Russian).

NUTTLI, O. W. (1985), *Average Seismic Source Parameter Relations for Plate-margin Earthquakes*, Tectonophys. *118*, 161–174.

PAPAGEORGIOU, A. S., and AKI, K. (1985), *Scaling Law of Far-field Spectra Based on the Observed Parameters of the Specific Barrier Model*, Pure Appl. Geophys. *123*, 353–374.

PURCARU, G., and BERCKHEMER, H. (1982), *Quantitative Relations of Seismic Source Parameters and Classification of Earthquakes*, Tectonophys. *84*, 57–128.

RAUTIAN, T. G., KHALTURIN, V. T., and ZAKIROV, M. S. (1981), *Experimental Studies of Seismic Coda*, Moscow, Nauka, 142 pp. (in Russian).

RIZNICHENKO, Yu. V. (Editor) (1960), *Methods of Detailed Studies of Seismicity*, Proc. Inst. Phys. Earth *9* (179) (in Russian).

SIPKIN, S. A. (1986), *Estimation of Earthquake Source Parameters by the Inversion of Wave Form Data: Global Seismicity 1981–1983*, Bull. Seismol. Soc. Am. *76*, 1515–1541.

SUGITO, M., and KAMEDA, H. (1985), *Prediction of Nonstationary Earthquake Motions on Rock Surface*, Proc. Jpn. Soc. Civ. Eng., Struct. Eng./Earthq. Eng. *2* (2), 149–159.

UTSU, T. (1982), *Relationships between Magnitude Scales*, Bull. Earthquake Res. Inst. *57*, 465–497.

VIKULIN, A. V. (1983), *On Relation between Energy and Magnitude Classification of Kamchatka, Kuriles and Japan*, Volcanol. and Seismol. *3*, 90–98.

YORK, D. (1966), *Least-squares Fitting of a Straight Line*, Canadian J. of Phys. *44*, 1079–1086.

ZHUO, T., and KANAMORI, H. (1987), *Regional Variation of the Short-period (1–10 Seconds) Source Spectrum*, Bull. Seismol. Soc. Am. *77*, 514–529.

(Received January 18, 1991, received/accepted August 14, 1991)

PAGEOPH, Vol. 136, No. 4 (1991)

0033−4553/91/040529−32$1.50 + 0.20/0

Complete Synthetic Seismograms for High-Frequency Multimode SH-waves

N. Florsch,[1,2] D. Fäh,[1,3] P. Suhadolc[1] and G. F. Panza[1]

Abstract—We present an efficient scheme to compute high-frequency seismograms (up to 10 Hz) for SH-waves in a horizontally stratified medium with the mode summation method. The formalism which permits the computation of eigenvalues, eigenfunctions and related integral quantities is discussed in detail. Anelasticity is included in the model by using the variational method. Phase velocity, group velocity, energy integral and attenuation spectra of a structure enable the computation of complete strong motion seismograms, which are the basic tool for the interpretation of near-source broad-band data.

Different examples computed for continental structures are discussed, where one example is the comparison between the observed transversal displacement recorded at station IVC for the November 4, Brawley 1976 earthquake and synthetic signals. In the case of a magnitude $M_L = 5.7$ earthquake in the Friuli seismic area we apply the mode summation method to infer from waveform modeling of all three components of motion of observed data some characteristics of the source.

Key words: Modal summation, broad band, Love waves, anelasticity.

1. Introduction

The mode summation method has been used (Swanger and Boore, 1978; Panza, 1985; Panza and Suhadolc, 1987) to model the response of a flat, layered earth since Thomson and Haskell (T-H) papers appeared (Thompson, 1950; Haskell, 1953). For the Rayleigh case, Knopoff (1964) proposed a modification of the initial T-H scheme which avoids a loss-of-precision intrinsic in the original formulation. This approach finally permitted the automatic computation of broadband synthetic seismograms for P-SV waves (e.g., Panza, 1985), which are complete in a given frequency-phase velocity window.

This paper is the expansion to SH-waves of the algorithm developed by Panza (1985) and Panza and Suhadolc (1987) for P-SV waves. The loss-of-precision

[1]Institute of Geodesy and Geophysics, University of Trieste, Via dell'Università 7, I-34100 Trieste, Italy.

[2]Currently at Laboratoire de Geophysique Applique, University of Paris 6, Paris, France.

[3]On leave from Institute of Geophysics, ETH, Zürich, Switzerland.

does not occur in the *SH*-wave case, while most of the other features of the *P-SV*-
and *SH*-computations are practically equivalent and transposable. A first-order
approximation of the anelasticity is applied including both Futterman's results
based on causality analysis (FUTTERMAN, 1962) and considerations on variational
methods (TAKEUCHI and SAITO, 1972; SCHWAB and KNOPOFF, 1972). This approach allows consideration of anelastic media characterized by Q as low as about
20. The attenuation effects obtained with this technique may be in error of about
0–20 percent in comparison with the exact method (SCHWAB, 1988; SCHWAB and
KNOPOFF, 1971, 1972, 1973).

The "mode-follower" procedure and structure minimization as described by
PANZA and SUHADOLC (1987) can be used in the *SH*-case. This approach permits
the calculation of "complete" synthetic seismograms with at least three significant
figures, as long as the distance to the source is greater than the wavelength (PANZA
et al., 1973). The seismograms computed in this way contain all the phases whose
phase velocities are smaller than the *S*-wave velocity of the halfspace terminating
the structural model.

The seismic source is introduced using BEN-MENAHEM and HARKRIDER'S
formalism (1964) while time duration is available through a convolutive model
(BEN-MENAHEM, 1961).

2. Computation of Eigenvalues

For the multimode surface-wave eigenvalue computations we make use of
SCHWAB and KNOPOFF's (1972) notation. The density-depth and velocity-depth
distributions in the earth are approximated with a structure composed of a series of
flat homogeneous layers. Then the dispersion function can be written as the
modified product for layer-matrices (SCHWAB and KNOPOFF, 1972):

$$F_L(\omega, c) = b_n \cdot b_{n-1} \cdot b_{n-2} \cdot \ldots b_1 \tag{1}$$

where n is the number of layers, including the lower halfspace. In equation (1) b_n
is given by:

$$b_n = (s, -1) \quad \text{if the halfspace is solid}$$
$$b_n = (0, -1) \quad \text{if the halfspace is liquid} \tag{2}$$
$$b_n = (1, 0) \quad \text{if the halfspace is rigid}$$

For the definition of the quantity s, see Eqs. (4).

The mathematical solution of the surface wave propagation allows two types of
waves in the solid halfspace, exponentially increasing and decreasing with depth. To
avoid infinite values of the solution, the coefficient of the exponentially increasing
wave in the halfspace must vanish.

If the halfspace is thought to be liquid, the deepest interface is at the analogy of the mantle-core boundary. Introduction of a rigid lower halfspace results in the locked mode approach (HARVEY, 1981). Then the halfspace becomes a perfect reflector and eigenvalues of the normal modes change by varying the depth of the halfspace.

b_m $(0 < m < n)$ is given by:

$$b_m = \begin{bmatrix} \cos Q_m & \dfrac{\sin Q_m}{\mu_m \cdot r_{\beta_m}} \\ \mu_m \cdot r_{\beta_m} \cdot \sin Q_m & \cos Q_m \end{bmatrix} \qquad \text{if} \quad c > \beta_m$$

$$b_m = \begin{bmatrix} \cosh Q_m^* & \dfrac{\sinh Q_m^*}{\mu_m \cdot r_{\beta_m}^*} \\ -\mu_m \cdot r_{\beta_m}^* \cdot \sinh Q_m^* & \cosh Q_m^* \end{bmatrix} \qquad \text{if} \quad c < \beta_m \qquad (3)$$

$$b_m = \begin{bmatrix} 1 & \dfrac{\omega \cdot d_m}{\mu_m \cdot c} \\ 0 & 1 \end{bmatrix} \qquad \text{if} \quad c = \beta_m.$$

In the expression (3)

$\mu_m = \rho_m \beta_m^2$ is the rigidity of the m-th layer,
β_m is the S-wave velocity of the m-th layer,
ρ_m is the density of the m-th layer,
d_m is the thickness of the m-th layer,
ω is the angular frequency,
c is the phase velocity.

Moreover,

$$\left. \begin{aligned} r_{\beta_m} &= \left(\left(\frac{c}{\beta_m} \right)^2 - 1 \right)^{1/2} \\ Q_m &= \frac{\omega \cdot r_{\beta_m} \cdot d_m}{c} = k \cdot r_{\beta_m} \cdot d_m \end{aligned} \right\} \quad \text{if} \quad c > \beta_m$$

$$\left. \begin{aligned} r_{\beta_m}^* &= -\left(1 - \left(\frac{c}{\beta_m} \right)^2 \right)^{1/2} \\ Q_m^* &= \frac{\omega \cdot r_{\beta_m}^* \cdot d_m}{c} = k \cdot r_{\beta_m}^* \cdot d_m \end{aligned} \right\} \quad \text{if} \quad c < \beta_m \qquad (4)$$

$$s = -\mu_n \cdot \left(1 - \left(\frac{c}{\beta_n} \right)^2 \right)^{1/2}$$

where k is the wavenumber.

The modified matrix product of b_m and b_{m-1} in equation (1) is defined as follows:

$$[b_m \cdot b_{m-1}]_{jp} = \begin{cases} (b_m)_{jl} \cdot (b_{m-1})_{lp} & \text{if } (j+p) \text{ is even} \\ (-1)^{j+1} \cdot (b_m)_{jl} \cdot (b_{m-1})_{lp} & \text{if } (j+p) \text{ is odd.} \end{cases}$$

Seeking eigenvalues (i.e., for a given phase velocity seeking the pulsation ω) requires the determination of the roots of the dispersion function. It can be done by root-bracketing and root-refining, according to a procedure described by SCHWAB and KNOPOFF (1972). This procedure is only necessary at the beginning of each mode. For all other points, the phase velocity can be estimated by cubic extrapolation and a root-refining procedure in the F-c plane (PANZA, 1985; PANZA and SUHADOLC, 1987).

Two kinds of overflow problems can occur. The first kind of overflow can appear when Q_m^* has a large absolute value. Then the calculation of $\cosh(Q_m^*)$ and $\sinh(Q_m^*)$ is prevented. In these cases, Q_m^* has always negative values and we can assume that the following approximations hold:

$$\cosh(Q_m^*) = \exp(-Q_m^*)$$

$$\sinh(Q_m^*) = -\exp(-Q_m^*).$$

In the matrix product (1) $\exp(-Q_m^*)$ can be factorized and finally set equal to one, since only the roots of the dispersion function are of interest. This operation also saves computation time. In analogy with the P-SV case (SCHWAB *et al.*, 1984), we call it the "single-layer" overflow control.

The second kind of overflow can appear when the whole matrix product (1) is computed with a phase velocity distant from the root. In this case a normalization procedure is used to prevent overflow: for each product the resulting 1×2 matrix is divided by the greatest absolute value of the matrix itself (SCHWAB and KNOPOFF, 1972). In analogy with the P-SV case (SCHWAB *et al.*, 1984), this is called the "multilayer" overflow control.

To handle realistic earth structural models, the computation scheme must allow for numerous layers in order to model possible gradients in the physical properties. Such gradients can be approximated by a sequence of thin layers. An optimized efficiency in the computations for such structures requires a mode follower and a structure minimization procedure, as described by PANZA and SUHADOLC (1987).

The structure minimization procedure is relevant to avoid computations in that part of the structure where the eigenfunction vanishes. It consists of an algorithm that keeps only the upper part of the structure for the computation, where the eigenfunction is not vanishing. This prevents a possible overflow in the calculation of eigenfunctions and saves computer time. Overflow can occur because the root is never exact and a residue remains in the exponentially increasing part of the downgoing wave.

To find the minimum of the eigenfunction above the part of the structure where overflow problems may occur, the function E_m is used. This quantity is defined as follows:

$$E_m = \rho_m \cdot \left(\frac{v_m}{v_0}\right)^2 \qquad (5)$$

where v_m is the displacement at the m-th interface, ρ_m is the density of the m-th layer and v_0 is the displacement at the surface.

The maximum depth of penetration of the considered mode corresponds to the deepest minimum of E_m. The layers below the minimum can be discarded, whereas the uppermost of them define the terminating halfspace.

Generally modes are very close to each other. This creates problems in following an individual mode in the phase velocity—frequency space and in distinguishing it from the neighbouring modes. The mode follower provides an efficient way to distinguish individual modes. It is based on the fact that for a given mode, the sign of $\partial F/\partial c$ is constant, whereas in going from a mode to the next sign changes. This condition, combined with possible values of the phase velocity at a given frequency, recognizes an eventual jump to a neighbouring mode.

3. Computation of Eigenfunctions

With the geometry shown in Figure 1, the computation of the eigenfunctions at the layer interfaces can be performed as follows (see e.g., SCHWAB, 1970):

$$\begin{bmatrix} v_m \\ (\sigma_z)_m \end{bmatrix} = \begin{bmatrix} \cos Q_m & \dfrac{\sin Q_m}{k \cdot \mu_m \cdot r_{\beta_m}} \\ -k \cdot \mu_m \cdot r_{\beta_m} \cdot \sin Q_m & \cos Q_m \end{bmatrix} \cdot \begin{bmatrix} v_{m-1} \\ (\sigma_z)_{m-1} \end{bmatrix} \quad \text{if} \quad c > \beta_m$$

$$\begin{bmatrix} v_m \\ (\sigma_z)_m \end{bmatrix} = \begin{bmatrix} \cosh Q_m^* & \dfrac{\sinh Q_m^*}{k \cdot \mu_m \cdot r_{\beta_m}^*} \\ k \cdot \mu_m \cdot r_{\beta_m}^* \cdot \sinh Q_m^* & \cosh Q_m^* \end{bmatrix} \cdot \begin{bmatrix} v_{m-1} \\ (\sigma_z)_{m-1} \end{bmatrix} \quad \text{if} \quad c < \beta_m \quad (6)$$

$$\begin{bmatrix} v_m \\ (\sigma_z)_m \end{bmatrix} = \begin{bmatrix} 1 & \dfrac{d_m}{\mu_m} \\ 0 & 1 \end{bmatrix} \cdot \begin{bmatrix} v_{m-1} \\ (\sigma_z)_{m-1} \end{bmatrix} \qquad \text{if} \quad c = \beta_m$$

where v_m is the displacement and $(\sigma_z)_m$ the stress at the interface m. Notice that:

$$\dot{v} = i\omega v.$$

These computations are performed using the initial values $(v_0, (\sigma_z)_0) = (1, 0)$ at the free surface.

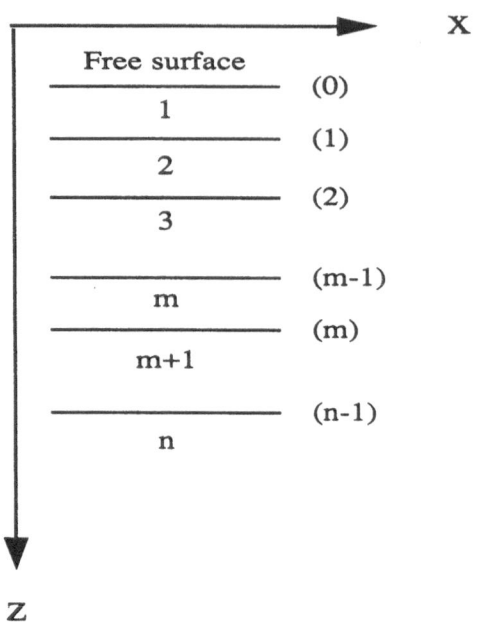

Figure 1
Coordinate system and geometry for the Love wave problem. The numbers denote layers, while the numbers in brackets denote interfaces.

As far as the phase velocity has been computed with high accuracy, the problem with "multilayer" overflow should not occur. However, "single-layer" overflow has to be expected because it depends on the thickness of the layer. By splitting each of the thickest layers in a series of equally thin layers this problem can be avoided.

4. Group Velocities

For a general background on the subject we refer to SCHWAB and KNOPOFF (1972). The group velocities are computed using the formula:

$$u = \frac{c}{1 - \frac{dc}{d\omega} \cdot \frac{\omega}{c}}. \tag{7}$$

The ratio $dc/d\omega$ is calculated according to the implicit function theory:

$$\frac{dc}{d\omega} = -\frac{\left[\dfrac{\partial F}{\partial \omega}\right]_c}{\left[\dfrac{\partial F}{\partial c}\right]_\omega}. \tag{8}$$

Since the S-wave velocities are frequency-dependent, all derivatives imply partial derivatives with respect to frequency. All required quantities are given in Appendix A.

Even if the dispersion function has been normalized to prevent "multilayer" overflow, it is not necessary to introduce the derivatives of the normalization coefficients to compute the group velocities. In fact the dispersion function without normalization is given by:

$$F_L(\omega, c) = \prod_{i=1}^{n} b_i(\omega, c)$$

where b_i is the matrix defined by (1), (2) and (3). This equation can be written in a recursive form (decreasing):

$$F_j = F_{j+1} \cdot b_j$$
$$F_n = b_n.$$

(9)

Normalization leads to the division of each b_i by a coefficient g_i, which is the greatest absolute value of the elements of the resulting 1×2 matrix F_i. Therefore the normalized dispersion function is written:

$$\hat{F}_L(\omega, c) = \prod_{i=1}^{n} \frac{b_i(\omega, c)}{g_i(\omega, c)}.$$

(10)

Hence:

$$\frac{\partial \hat{F}_L}{\partial c} = \frac{\partial F_L}{\partial c} \cdot \prod_{i=1}^{n} \frac{1}{g_i} + F_L \cdot \frac{\partial}{\partial c} \left(\prod_{i=1}^{n} \frac{1}{g_i} \right).$$

A similar relation holds for $\partial \hat{F}_L / \partial \omega$. Therefore, we obtain:

$$\frac{dc}{d\omega} = -\frac{\left[\dfrac{\partial \hat{F}_L}{\partial \omega} \right]_c}{\left[\dfrac{\partial \hat{F}_L}{\partial c} \right]_\omega} = -\frac{\left[\dfrac{\partial F_L}{\partial \omega} \right]_c \cdot \prod_{i=1}^{n} \dfrac{1}{g_i} + F_L \cdot \dfrac{\partial}{\partial \omega} \left[\prod_{i=1}^{n} \dfrac{1}{g_i} \right]}{\left[\dfrac{\partial F_L}{\partial c} \right]_\omega \cdot \prod_{i=1}^{n} \dfrac{1}{g_i} + F_L \cdot \dfrac{\partial}{\partial c} \left[\prod_{i=1}^{n} \dfrac{1}{g_i} \right]}.$$

(11)

If the phase velocity has been computed with high accuracy, then F_L is very close to zero and (11) reduces to:

$$\frac{dc}{d\omega} = -\frac{\left[\dfrac{\partial \hat{F}_L}{\partial \omega} \right]_c}{\left[\dfrac{\partial \hat{F}_L}{\partial c} \right]_\omega} = -\frac{\left[\dfrac{\partial F_L}{\partial \omega} \right]_c \cdot \prod_{i=1}^{n} \dfrac{1}{g_i}}{\left[\dfrac{\partial F_L}{\partial c} \right]_\omega \cdot \prod_{i=1}^{n} \dfrac{1}{g_i}} = -\frac{\left[\dfrac{\partial F_L}{\partial \omega} \right]_c}{\left[\dfrac{\partial F_L}{\partial c} \right]_\omega}.$$

(12)

This equation does not hold if the derivative $\partial F / \partial c$ is computed away to the root of F, as in the case, where $\partial F / \partial c$ is used in the root-refining procedure of the dispersion function. In this situation a recursive approach has been chosen.

Equation (10) can be written in a recursive form as eqution (9):

$$\hat{F}_j = \frac{\hat{F}_{j+1} \cdot b_j}{g_j}$$

$$\hat{F}_n = b_n$$

where b_n and $\hat{F}_j$ are 1×2 matrices, b_j is a 2×2 matrix and g_j is a scalar value. The normalization has not been performed for the halfspace ($j = n$). The derivative of the normalized dispersion-function can now be determined with the same recursive scheme:

$$\frac{\partial \hat{F}_j}{\partial c} = \frac{\partial \hat{F}_{j+1}}{\partial c} \cdot \frac{b_j}{g_j} + \hat{F}_{j+1} \cdot \frac{\partial}{\partial c}\left(\frac{b_j}{g_j}\right)$$

$$= \frac{\partial \hat{F}_{j+1}}{\partial c} \cdot \frac{b_j}{g_j} + \hat{F}_{j+1} \cdot \frac{\dfrac{\partial b_j}{\partial c} \cdot g_j - b_j \cdot \dfrac{\partial g_j}{\partial c}}{g_j^2}$$

$$\frac{\partial \hat{F}_n}{\partial c} = \frac{\partial b_n}{\partial c}.$$

5. Energy Integral

This additional quantity is necessary for the computation of seismograms (PANZA and SUHADOLC, 1987). The energy integral is defined as:

$$I_1 = \int_0^\infty \rho \cdot \left(\frac{\dot{v}(z)}{\dot{v}_0}\right)^2 dz. \tag{13}$$

The energy integral can be calculated analytically, since simple analytic expressions are known for the eigenfunction $v(z)$. The details of these calculations are given in Appendix B.

6. Attenuation Due to Anelasticity

The treatment of anelasticity requires, for causality reasons, the introduction of body wave dispersion (FUTTERMAN, 1962). In a medium with constant Q, the *SH* phase velocity can be expressed as:

$$B_1(\omega) = \frac{B_1(\omega_0)}{1 + \dfrac{2}{\pi} \cdot B_1(\omega_0) \cdot B_2(\omega_0) \cdot \ln\left(\dfrac{\omega_0}{\omega}\right)}. \tag{14}$$

The layer index m is omitted. $B_1(\omega_0)$ and $\beta_2(\omega_0)$ are the S-wave velocity and the S-wave phase attenuation at the reference angular frequency ω_0. The quantities B_1 and B_2 are related to the complex body-wave velocity β (SCHWAB and KNOPOFF, 1972):

$$\frac{1}{\beta} = \frac{1}{B_1} - i \cdot B_2.$$

In the computation we have chosen the reference angular frequency $\omega_0 = 2\pi$ radians. In anelastic media the surface wave phase velocity c must be also expressed as a complex quantity:

$$\frac{1}{c} = \frac{1}{C_1} - i \cdot C_2$$

with C_1 the attenuated phase velocity and C_2 the phase attenuation, the latter being necessary for the computation of seismograms. C_2 can be estimated by using the variational technique (TAKEUCHI and SAITO, 1972; AKI and RICHARDS, 1980). The phase attenuation C_2 is given by:

$$C_2 = \frac{\displaystyle\int_0^\infty \mu \cdot B_1 \cdot B_2 \cdot \left(\frac{\sigma_z^2}{\mu^2 \cdot k^2} + v^2 \right) dz}{\displaystyle c \int_0^\infty \mu \cdot v^2 \, dz}. \tag{15}$$

This integral can be calculated analytically, since simple analytic expressions are known for the eigenfunctions. The details of this computation are given in Appendix B.

The most important effect of the attenuation is the modification of the wave velocities and the decay of amplitude in the final computations of seismograms. As the variational technique is only an approximate method, the Q values can be in error by as much as $0-20$ percent in comparison with the exact method. This error arises mainly from the use of the elastic and, therefore, real eigenfunctions to compute the phase attenuation.

Recently DAY *et al.* (1989) showed the limits of the variational technique in the locked mode approximation, which can be obtained by limiting the model with a rigid or liquid halfspace. They showed that an error in amplitudes up to 100 percent can occur, when dealing with low Q-values. The error increases when the Q-values undergo large variations with depth. Introducing a solid halfspace in the model and using the structure minimization procedure prevents this kind of error.

7. Source

In order to include the seismic source in the computations, the formulation due to HARKRIDER (1970) and BEN-MENAHEM and HARKRIDER (1964) is used. For the double-couple point source, the Fourier transform of the response can be written for a given mode as:

$$U_v = |\vec{n}| \cdot |R(\omega)| \cdot e^{i\Phi_0} \cdot e^{-i3\pi/4} \cdot k^{1/2} \cdot \chi(\theta, h) \cdot A_L \cdot \frac{e^{-ikr}}{\sqrt{2\pi r}} \cdot e^{-\omega r C_2} \qquad (16)$$

where $R(\omega)$ is the Fourier transform of the source time function and $\Phi_0 = \arg(R(\omega))$ is the source apparent initial phase. $|\vec{n}|$ is the absolute value of the normal vector to the plane of motion, with units of length. The factor A_L is given by:

$$A_L = \frac{1}{2 \cdot c \cdot u \cdot I_1}.$$

The effect of anelasticity is expressed by the term:

$$e^{-\omega r C_2}$$

$\chi(\theta, h)$ is the azimuthal dependence given by:

$$\chi(\theta, h) = i \cdot (d_1 \sin \theta + d_2 \cos \theta) + d_3 \sin 2\theta + d_4 \cos 2\theta$$

for a double-couple source,

$$d_1 = G(h) \cdot \cos \lambda \cdot \cos \delta$$

$$d_2 = -G(h) \cdot \sin \lambda \cdot \cos 2\delta$$

$$d_3 = \tfrac{1}{2} \cdot V(h) \cdot \sin \lambda \cdot \sin 2\delta$$

$$d_4 = V(h) \cdot \cos \lambda \cdot \sin \delta$$

θ is the angle between the strike of the fault and the epicenter-station direction, λ is the rake angle, δ is the dip angle and h is the source depth. The source geometry

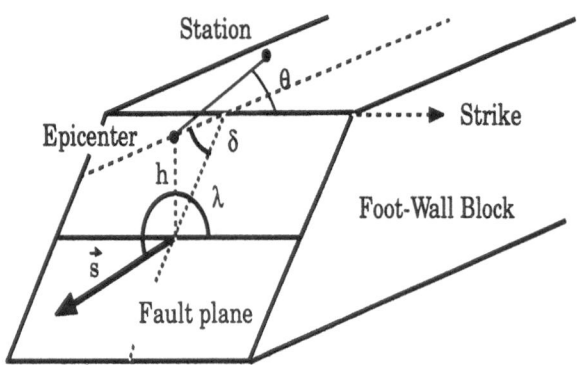

Figure 2

Source geometry and coordinate system associated with the free surface. θ is the angle between the strike of the fault and the epicenter-station direction, δ is the dip, λ is the rake and h is the source depth.

and the coordinate system associated with the free surface is given in Figure 2. $G(h)$ and $V(h)$ depend on the values of the eigenfunctions at the hypocenter:

$$G(h) = +\frac{1}{\mu_s} \cdot \left(\frac{\sigma_s^*(h)}{\frac{\dot{v}_0}{c}}\right) = \frac{1}{k \cdot \mu_s} \cdot \frac{\sigma_s(h)}{v_0}$$

$$V(h) = \frac{\dot{v}_s(h)}{\dot{v}_0} = \frac{v_s(h)}{v_0}.$$

v_0 is the value of the eigenfunction at the surface and $\sigma_s(h)$ is the stress at the depth of the source. Equation (16) is equivalent to equation (7.148) given in AKI and RICHARDS (1980). The seismogram related to a given mode is obtained by the inverse Fourier transform of (16).

8. Examples of Computation

Frequency Domain

The layered model in Table 1 represents an average structure of the Friuli seismic area in the southern pre-Alps, close to the May 6, 1976, Friuli earthquake. The same structure was used to illustrate the mode summation for Rayleigh-waves (PANZA and SUHADOLC, 1987).

Table 1

Structure FRIUL7A. Q_α is taken as $2.5Q_\beta$

Thickness [km]	Density [g/cm³]	P-wave velocity [km/s]	S-wave velocity [km/s]	Q_β
0.04	2.00	1.50	0.60	20
0.06	2.30	3.50	1.80	30
0.20	2.40	4.50	2.50	100
0.70	2.40	5.00	2.90	200
2.00	2.60	6.00	3.30	400
3.50	2.60	6.20	3.45	400
4.50	2.60	6.00	3.35	100
10.00	2.60	5.50	3.30	50
3.50	2.60	6.00	3.50	100
2.50	2.75	6.50	3.75	400
2.50	2.80	7.00	3.85	400
7.50	2.80	6.50	3.75	100
4.00	2.85	7.00	3.85	200
3.00	3.20	7.50	4.25	400
1.50	3.40	8.00	4.50	400
9.00	3.45	8.20	4.65	400

1. Phase Velocities

The dispersion curves for the first 154 Love modes are shown in Figure 3. For
S-wave velocities less than 3.35 km/s the modes are well separated. This velocity
corresponds to the S-wave velocity in the upper part of the crustal low-velocity
zone (LVZ). Modes situated in the part of the spectrum below this phase velocity
value, sample therefore the part of the crust above the uppermost LVZ.

In the part of the spectrum with higher phase velocities the dispersion curves are
packed together. An enlarged portion of this part is presented in Figure 3b. Since
two LVZ are present in the structural model, areas are seen where the higher Love
wave modes decompose into families of low-velocity channel waves and families of
waves propagating in the upper crust. They appear in the dispersion curves as an
apparent continuity of the phase velocities between adjacent modes. This mode-to-
mode continuation leads to the identification of a family of waves. Each member of
a wave family begins with one of the Love-wave modes and contains segments of
all successive higher modes. They have almost continuous phase velocities, broken
only at the points of near-oscillations. The segments of members of the family of
upper-crustal waves form apparently continuous curves which sometimes seem to
intersect the more horizontal trending family of the channel-wave curves. A

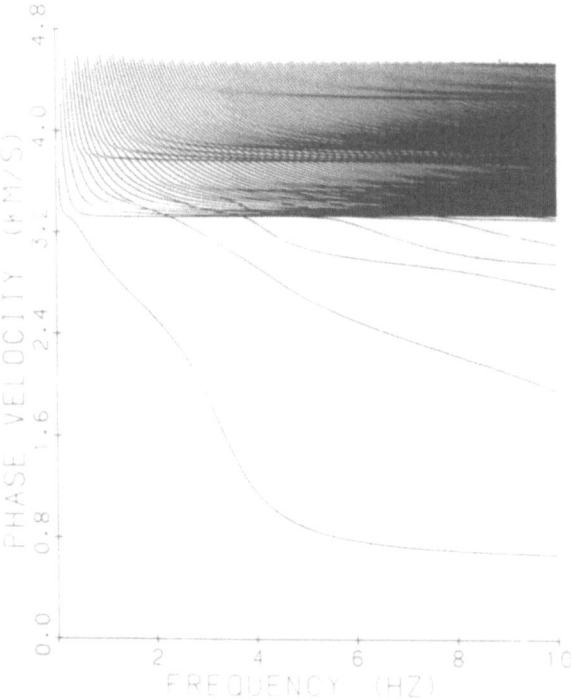

Figure 3a

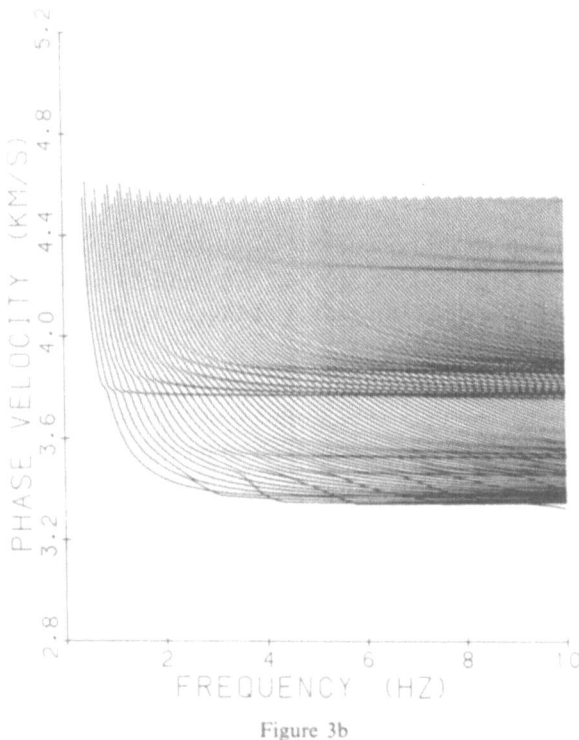

Figure 3b

Figure 3

(a) Love-wave dispersion curves for the structural model FRIUL7A. The mode numbering is the following: 0 for the fundamental mode, 1 for the first higher mode, 2 for the second higher mode, and so on up to 153. (b) Enlarged portion (modes 6–153) of part (a) showing the effect of low-velocity waveguides.

member of the family of upper-crustal waves can be identified at a frequency of about 4 Hz in the phase velocity range 3.35–3.45 km/s.

Another type of apparent continuity of the phase velocities for adjacent modes can be related to the structural layering (for example at a phase velocity of about 4.25 km/s). Such parts of the spectrum represent refracted waves at strong elastic impedance contrasts. They are characterized by phase velocities which tend to become constant with increasing frequency.

2. Group Velocities

The group velocity spectrum is presented in Figure 4. Due to the complexity of the pattern, it has been divided into two parts. Modes with group velocities less than about 2.8 km/s correspond to waves propagating in the low-velocity sediments.

In the part of the spectrum where group velocities are in the interval 2.8–3.2 km/s, several higher modes form stationary phases. They correspond to families of waves propagating in the upper crust and are characterized by the same type of mode-to-mode continuation as in the phase velocity curves. They can be interpreted as the high-frequency equivalent of *Lg* phases (SCHWAB and KNOPOFF, 1972; KNOPOFF *et al.*, 1973; PANZA and CALCAGNILE, 1975), which are propagating in the upper part of the continental crust.

The flat portions of group velocity curves formed by a large number of higher modes at about 3.35 km/s (Figure 4a) and 3.75 km/s (Figure 4b) correspond to waves propagating in the upper and lower channel.

3. Energy Integral

The energy integral can serve as an estimate of the contribution of the different modes to the surface displacement. In general, neglecting the influence of the source depth on the excitation of different modes, small values of the energy integral I_1 correspond to large surface displacements. In the whole frequency range, the fundamental mode has the lowest values of I_1 (Figure 5). For a shallow source, the fundamental mode generally dominates the surface displacement.

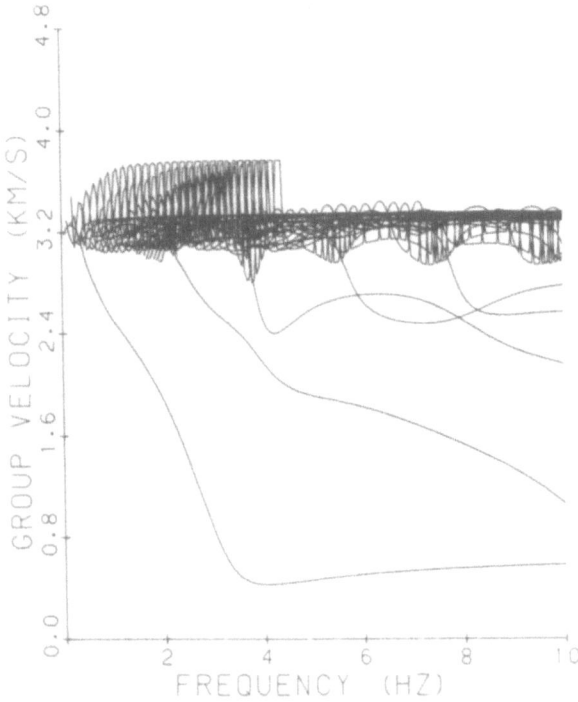

Figure 4a

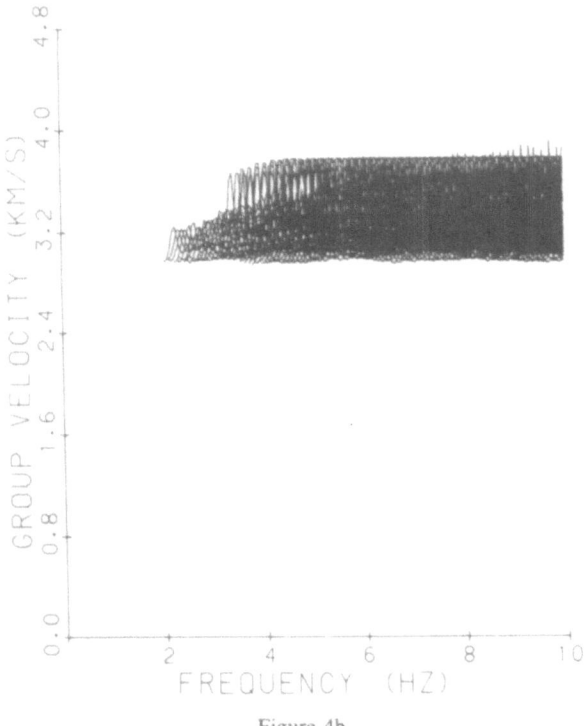

Figure 4
Love-wave group velocities for the structure FRIUL7A. The spectrum is divided into two parts: a) Love
modes 0–30, b) Love modes 31–153.

The mode-to-mode continuations in the lower part of the energy integral curves
(Figure 5a) correspond to the high frequency equivalent of *Lg* waves. The low
values of the energy integral indicate that these waves can give rise to significant
amplitudes at the surface.

Most of the energy of channel waves is concentrated in the channel. Therefore,
the energy integral of these families, seen in the upper part of Figure 5a, takes
higher values than those for upper-crustal waves. For a given member of this family
the maximum displacement in the low-velocity zone becomes larger, relative to the
displacement at the free surface, with increasing frequency. Therefore, the energy
integral of this member is characterized by values increasing with frequency. This
can be seen in the general pattern of the upper part of Figure 5a.

4. Quality Factor Q_x

The phase attenuation C_2 of the *SH*-modes is related to the quality factor Q_x by
the relation

$$1/Q_x = 2C_1 C_2$$

where C_l is the anelastic phase velocity (SCHWAB and KNOPOFF, 1972). The quality factor is presented in Figure 6. Q_β is very low in the sedimentary layers. Modes mainly propagating in these layers are therefore characterized by low Q_x values (Figure 6a). This is the case for the first few modes, especially for the fundamental and first higher mode. The effect of layering of Q_β can be observed for several nearby modes that have almost constant Q_x, for example Q_x close to 65. The resulting Q_x values are close to the values Q_β of the structural model for those Love wave modes, whose eigenfunction mainly sample the corresponding part of the structure.

Time Domain

The first example corresponds to the November 4, 1976 Brawley, California earthquake. The structural model and source parameters have been proposed by HEATON and HELMBERGER (1978). Their structural model is given in Table 2. Since SWANGER and BOORE (1978) computed synthetic seismograms for this event with the mode-summation technique, their result, even if limited to the elastic case, provides a test for our programs. Therefore, the source parameters used to compute synthetics are the same as those given by SWANGER and BOORE.

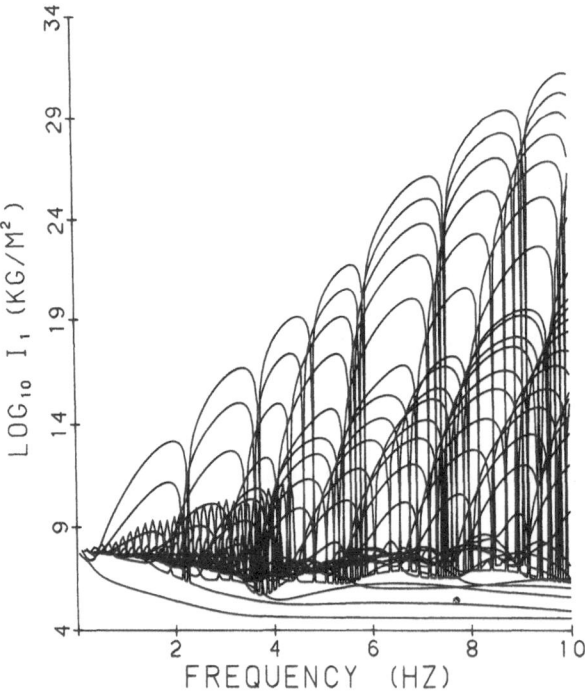

Figure 5a

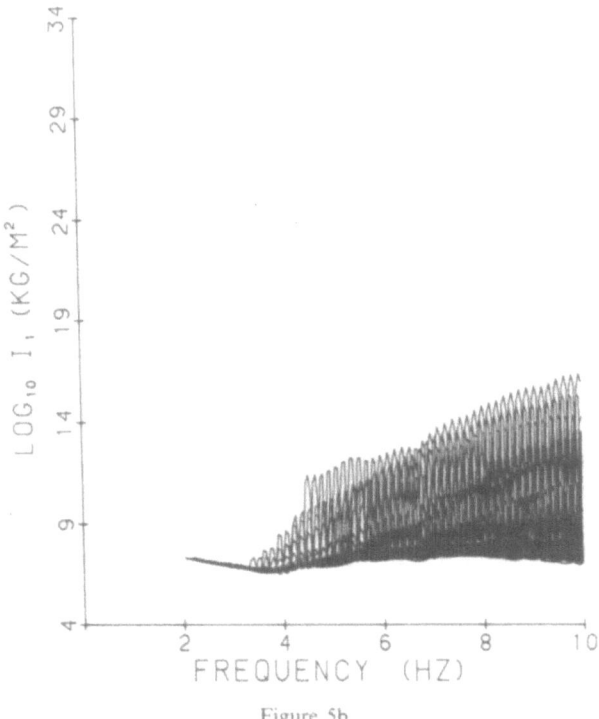

Figure 5b

Figure 5

Love-wave energy integral I_1 for the structure FRIUL7A. The spectrum is divided into two parts: a) Love modes 0–30, b) Love modes 31–153.

A strike-slip point source is placed on a vertical plane at 6.9 km depth. The rupture-velocity time-function is a symmetrical triangle with a base of 1.5 seconds. At a distance of 33 km from the source, the displacement consists almost entirely of the fundamental mode and the first higher modes (Figure 7). The recorded displacement at the station IVC, 33 km from the source, is given in the same figure. The upper frequency limit is 1 Hz. It can be seen that there is generally a very good agreement between the two synthetic signals.

Table 2

Imperial Valley Structure proposed by HEATON *and* HELMBERGER *(1978)*

Layer	Thickness [km]	Density [g cm^{-3}]	S-velocity [km s^{-1}]
1	0.95	1.80	0.88
2	1.15	2.35	1.50
3	3.80	2.60	2.40
halfspace	∞	2.80	3.70

In the second example we present synthetic seismograms for the structural model FRIUL7A shown in Table 1. The upper frequency limit is 10 Hz. The source parameters are related to those of the Friuli, May 6, 1976 earthquake (point-source approximation with the source parameters taken from SUHADOLC *et al.*, 1988). The receivers are chosen in the direction of the dominant lobe of the radiation pattern of *SH*-waves (north-east direction, with a strike-receiver angle of 235°), resulting from the selected source parameters. Synthetic ground displacements, velocities and accelerations are presented in the lower part of Figure 8. The signals are filtered with a Gaussian filter (the first filtered frequency is at 9 Hz, with a reduction of the amplitude by factor 100 at the cutoff frequency of 10 Hz). This filter prevents ringing due to the cutoff frequency. A decomposition of the displacement into different sets of modes is presented in the upper part of the figure. It shows that the higher modes are essential in defining the shape of the waveform, especially in the body wave part of the synthetic seismograms.

In Figure 9 synthetics due to a source with finite rise time are presented. The rupture-velocity time-function is a symmetrical triangle with a base of 0.5 seconds. The signals are filtered with the already described Gaussian filter. As expected, the energy is shifted to lower frequencies, as the duration of the source is increased. The

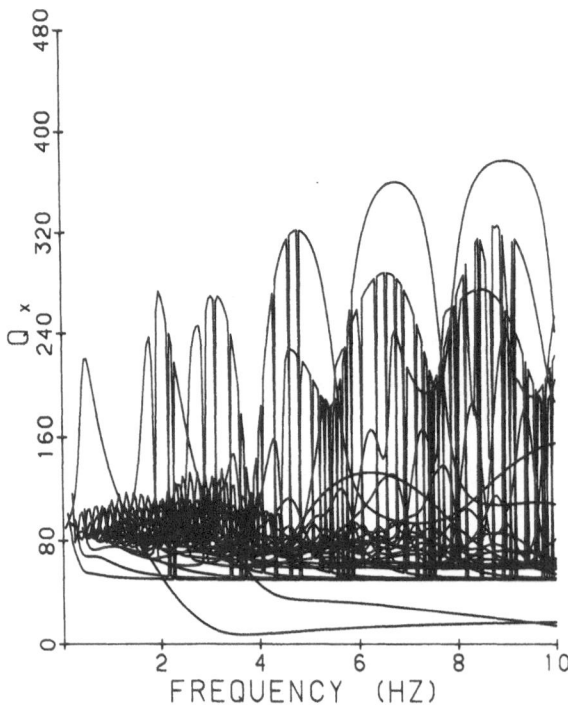

Figure 6a

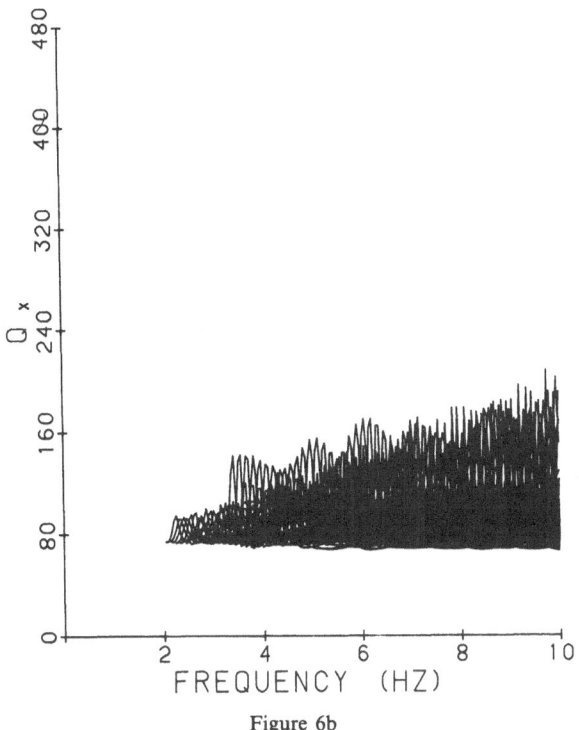

Figure 6b

Figure 6

Love-wave quality factor Q_x for the structure FRIUL7A. The spectrum is divided into two parts: a) Love Modes 0–30, b) Love modes 31–153.

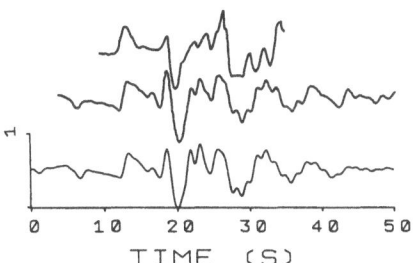

Figure 7

Comparison between the observed ground displacement (top trace), the synthetic signals (middle trace) computed by SWANGER and BOORE (1978) and our synthetics (lowest trace) for the Brawley, 1976 earthquake as recorded at station IVC. For the synthetic signals a vertical right-lateral strike-slip point source with duration of 1.5 s, placed on a vertical plane at 6.9 km depth, is considered. All amplitudes are normalized to a source with a seismic moment of 1 dyne cm. The peak displacement is $6.0 \cdot 10^{-25}$ cm.

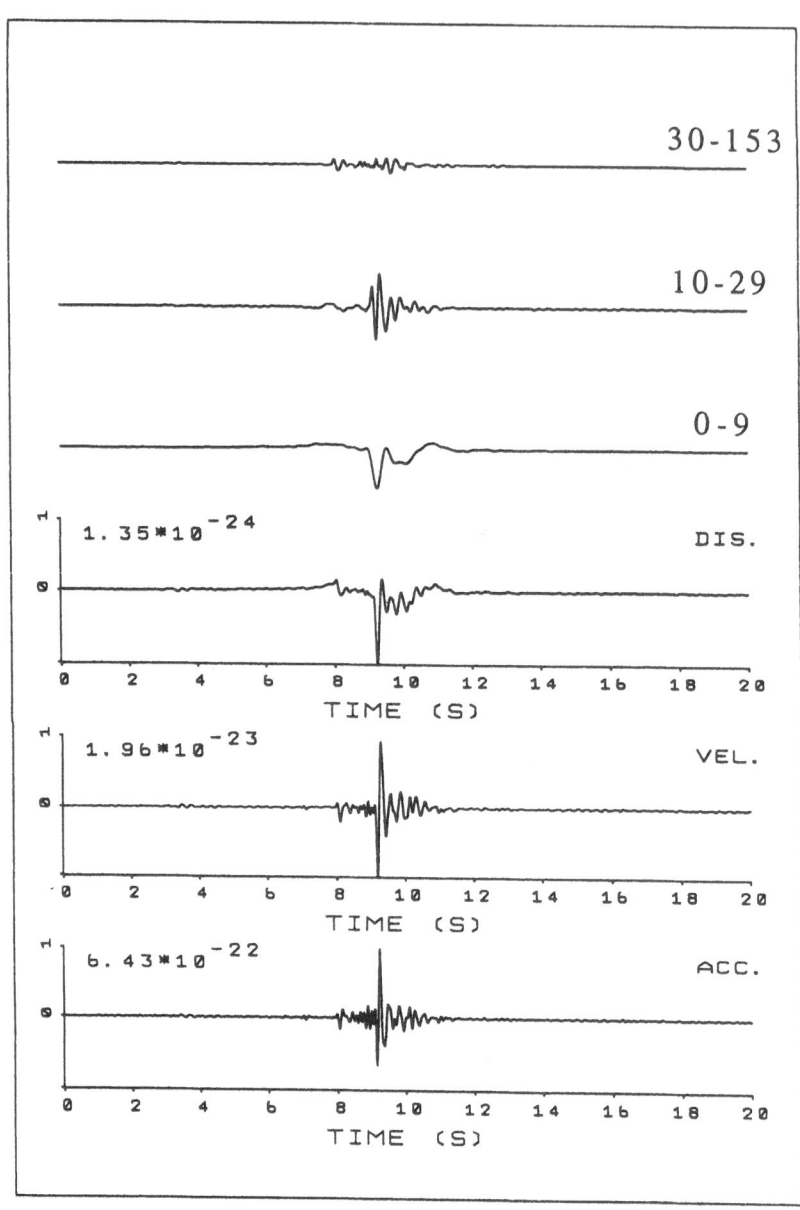

Figure 8
Displacement, velocity and acceleration (lower three traces), computed for a receiver placed 30 km from the source. The displacement is decomposed in different sets of modes (upper three traces: Love modes 30–153, Love modes 10–29, Love modes 0–9). It shows the contribution of the higher modes to the signal waveform. An instantaneous point source with a depth of 7 km is considered (angle strike-receiver $\phi = 280°$, dip $\delta = 30°$ and rake $\lambda = 115°$). All amplitudes are normalized to a source with seismic moment of 1 dyne cm. The peak displacement is in units of cm, the peak velocity in units of cm s^{-1} and the peak acceleration in units of cm s^{-2}.

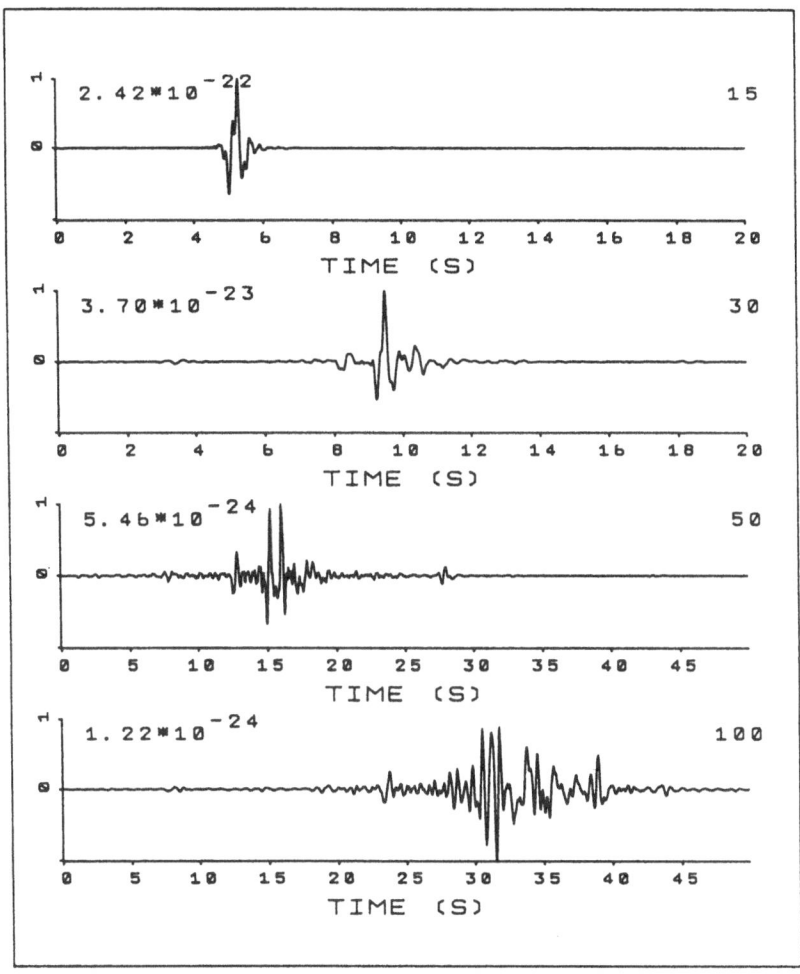

Figure 9

Acceleration time series at different distances from the source (15 km, 30 km, 50 km and 100 km). A point source with duration of 0.5 s and 7 km depth is considered (angle strike-receiver $\phi = 280°$, dip $\delta = 30°$ and rake $\lambda = 115°$). All amplitudes are normalized to a source with seismic moment of 1 dyne cm. The peak acceleration is in units of cm s^{-2}.

strong phases at about 35 s, for the signal at 100 km distance from the source (lowest trace), can be identified as the *Lg* wavetrain.

The last example compares synthetic seismograms and observed data for the September 11, Friuli (Italy) aftershock (16:35). The event has been recorded by various accelerograph stations (CNEN-ENEL, 1977) The three component uncorrected seismograms recorded at the station Buia are shown in Figure 10a.

PANZA and SUHADOLC (1987) have shown, assuming the 1-D layered, anelastic structural model FRIUL7A, that the observed vertical signal at the station Buia cannot be explained by one point source only. A good fit was obtained with three point sources, having different weights and time shifts, but the same focal depth and mechanism. The same conclusion was drawn by MAO *et al.* (1990) by modeling all three components of the recorded seismograms. In Figure 10b, the results obtained by MAO *et al.* (1990) are shown, where *SH*- and *P-SV*-waves have been combined. The layered *P*- and *S*-wave velocity model (FRIUL7W in Table 3) used in their study is slightly different from the model FRIUL7A. It is based on the result of a damped least-square inversion of arrival time data from local earthquakes (MAO and SUHADOLC, 1990). The differences between FRIUL7A and FRIUL7W are the depth and shape of the upper low-velocity zone, the depth of the sedimentary cover and the quality factors.

To fit the observed seismograms, the source is approximated by a sum of point sources using a trial-and-error technique. The parameters varied in the process are the number of point sources, their origin time and the weights of the single sources. The distance to the receiver, the source depth, the strike, dip and rake are varied, but kept constant for all subevents. All these parameters are adjusted until satisfactory (in the least-square sense) waveform fit was obtained, both in the time and in the frequency domain.

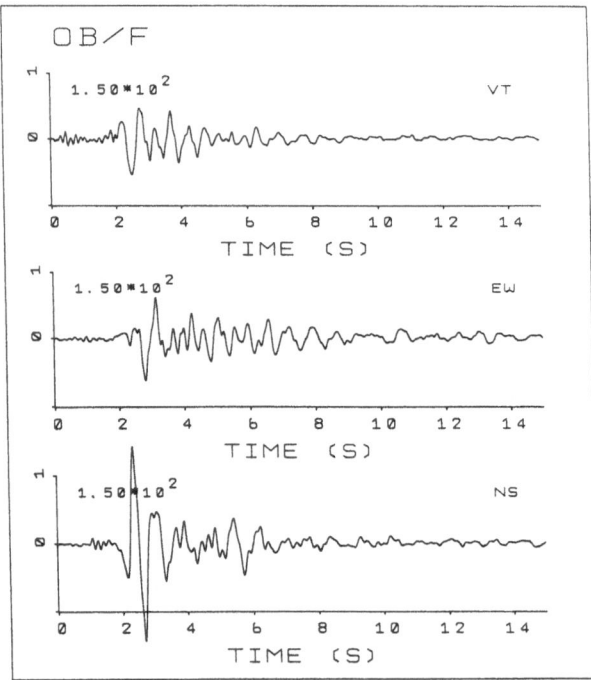

Figure 10a

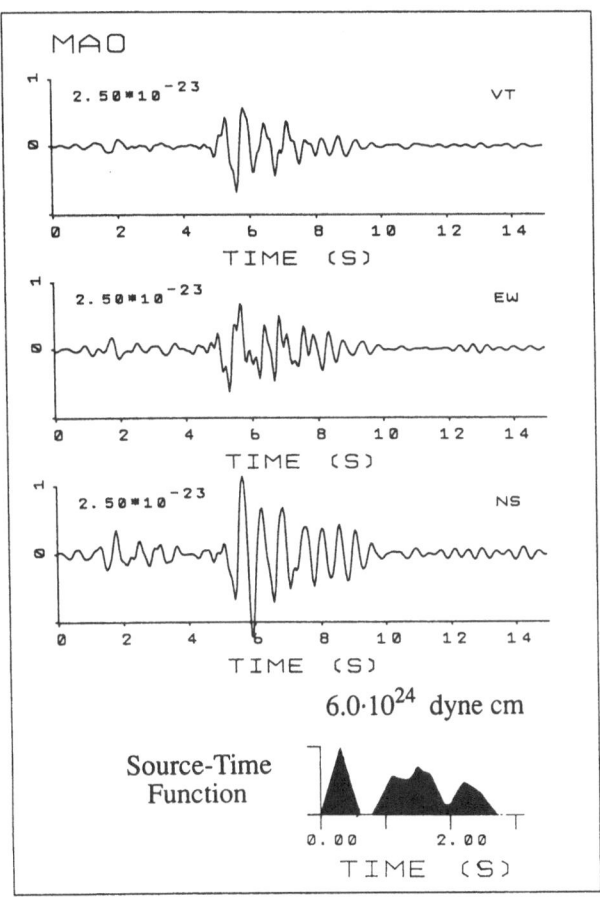

Figure 10b

Figure 10

Comparison of the observed ground motion at the station Buia for the September 11, Friuli aftershock (16:35) with results from waveform fitting with the mode summation technique for a layered, anelastic structure. a) Uncorrected accelerograms, after Gaussian filtering, with a cutoff frequency of 7.5 Hz, for the September 11, Friuli $M_L = 5.7$ aftershock (16:35), observed at the station Buia. The zero of the time axes does not coincide with the origin time. The amplitudes are given in cm s^{-2}. b) Synthetic accelerograms corresponding to six seismic point sources located at the same depth of 17.1 km and the same distance of 15 km (MAO *et al.*, 1990). The strike, the angle between the strike of the source and the receiver, the dip and the rake are 225°, 19°, 28° and 115°, respectively. The six point sources have different weights and time shifts (1.0, 0.6, 0.6, 0.6, 0.5, 0.5 and 0 sec, 0.77 sec, 1.13 sec, 1.37 sec, 1.9 sec, 2.18 sec). In the lowest part of the figure, the normalized source-time function corresponding to the synthetic signals is shown. The seismic moment of 6.0 · 10^{24} dyne cm corresponds to the value which gives the best fit of the synthetic to the observed signals.

Table 3

Structure FRIUL7A. *(MAO and SUHADOLC, 1990). Q_α is taken as $2.5Q_\beta$*

Thickness [km]	Density [g/cm³]	P-wave velocity [km/s]	S-wave velocity [km/s]	Q_β
0.057	2.00	1.50	0.60	20
0.043	2.30	3.50	1.80	20
0.20	2.40	4.50	2.50	50
0.70	2.40	5.55	3.05	100
2.00	2.60	5.88	3.24	100
0.10	2.60	5.70	3.14	50
0.20	2.60	5.65	3.10	50
0.20	2.60	5.60	3.06	50
1.00	2.60	5.57	3.03	50
0.50	2.60	5.55	3.02	50
1.00	2.60	5.57	3.03	50
0.20	2.60	5.60	3.06	50
0.20	2.60	5.65	3.10	50
0.10	2.60	5.70	3.14	50
4.50	2.60	5.88	3.25	100
0.10	2.60	6.10	3.40	200
0.10	2.60	6.20	3.50	200
0.10	2.60	6.30	3.60	200
0.70	2.60	6.45	3.75	200
2.50	2.60	6.47	3.77	200
5.00	2.60	6.50	3.80	200
5.00	2.60	6.55	3.82	200
1.00	2.75	6.55	3.82	200
2.00	2.75	7.00	3.85	200
2.00	2.80	7.00	3.85	200
7.50	2.80	6.50	3.75	100
4.00	2.85	7.00	3.85	200
3.00	3.20	7.50	4.25	400
1.50	3.40	8.00	4.50	400
9.00	3.45	8.20	4.65	400

In all cases of waveform fitting, the orientation of the sources agree well with previously published results by SLEJKO and RENNER (1984), who interpreted the event as a thrust on a very shallow NW dipping plane. To fit the observed signals, several point sources with different weights and time shifts are required. The vertical component and the first phases of the seismograms can be well reproduced, but the NS-component of the synthetic seismograms has too big amplitudes in the coda. The duration of the observed EW component cannot be reproduced with this set of point sources. The station Buia was placed in a sedimentary basin and effects like local surface waves in the basin or two-dimensional resonances can become important. Lateral heterogeneities could therefore account for the differences between the observed and the computed signals for the two horizontal components.

The vertical component of motion is less sensitive to such site-effects, as for example observed at different sites in Mexico City during the 1985 Michoacan earthquake (CAMPILLO *et al.*, 1988).Therefore, considerably better results are obtained for this component.

9. Conclusions

In the case of a layered structure, the mode summation method is a powerful tool to compute synthetic broad-band seismograms. The mode follower and structure minimization allow inclusion of low-velocity zones. Such computations are very efficient and stable. The resulting seismograms include first-order effects due to anelasticity, e.g., the intrinsic attenuation and body-wave dispersion. One of the most attractive aspects of the presented phase and group velocity spectrums is the possibility to identify particular phase arrivals.

One may wonder whether the proposed method is applicable to local structures, which have seldom plane-layer characteristics. In fact, high-frequency seismograms are very sensitive to lateral heterogeneities. Their influence should be included in the numerical modeling. This in turn, requires the use of at least 2-D models to take into account different tectonic settings and site effects. One elegant and efficient way to handle this problem is either by the 2-D mode summation method (VACCARI *et al.*, 1989) for different tectonic settings, or through the combined use of the modal summation method and the finite difference technique (FÄH *et al.*, 1990) for treating site effects.

The mode summation method presented in this paper can be applied in many fields, especially in broad-band studies to analyze recorded regional earthquakes. It can serve to predict radiation patterns of Love waves and can be used in seismic hazard assessments.

Appendix A: Derivatives of the Matrices for the Computation of the Group Velocities

Three cases have to be distinguished: $c > \beta_m$, $c < \beta_m$ and $c = \beta_m$. To avoid excessively heavy notations, layer indexes m and n are dropped. The *S*-wave velocity is denoted by β. Introducing body-wave dispersion, we have:

$$\beta = \frac{B_1(\omega_0)}{1 + \frac{2}{\pi} \cdot B_1(\omega_0) \cdot B_2(\omega_0) \cdot \ln\left(\frac{\omega_0}{\omega}\right)}.$$

$B_1(\omega_0)$ and $B_2(\omega_0)$ are, respectively, the phase velocity and the phase attenuation at the reference angular frequency ω_0. The rigidity is $\mu = \rho\beta^2$ and ρ is the density. In the following, d is the layer thickness and k is the wavenumber. Derivatives with respect to c denoted using a dot symbol ($\cdot$), while those with respect to ω are denoted with a prime symbol ($'$).

If the halfspace is solid, we use the quantity s (SCHWAB and KNOPOFF, 1972):

$$s = -\mu \cdot \left(1 - \left(\frac{c}{\beta}\right)^2\right)^{1/2}$$

$$\dot{s} = \frac{\partial s}{\partial c} = \frac{\mu \cdot c}{\beta^2 \cdot \left(1 - \left(\frac{c}{\beta}\right)^2\right)^{1/2}}$$

$$s' = \frac{\partial s}{\partial \omega} = \frac{2 \cdot s \cdot B_2(\omega_0) \cdot \beta}{\pi \cdot \omega} \cdot \left(2 + \frac{c^2}{\beta^2 - c^2}\right).$$

If the halfspace is liquid

$$s = \dot{s} = s' = 0$$

while if the halfspace is rigid

$$s = 1 \quad \text{and} \quad \dot{s} = s' = 0.$$

First Case: $c > \beta$

We use here the following notations:

$$r = \left(\left(\frac{c}{\beta}\right)^2 - 1\right)^{1/2}$$

$$Q = \omega \cdot r \cdot \frac{d}{c} = k \cdot r \cdot d.$$

The layer matrix is:

$$b = \begin{bmatrix} \cos Q & \dfrac{\sin Q}{\mu \cdot r} \\ \mu \cdot r \cdot \sin Q & \cos Q \end{bmatrix}.$$

We have

$$\dot{b}_{11} = -\frac{Q \cdot \beta^2 \cdot \sin Q}{c \cdot (c^2 - \beta^2)}$$

$$\dot{b}_{12} = \frac{1}{\mu} \cdot \left(-\frac{c \cdot \sin Q}{\beta^2 \cdot r^3} + \frac{Q \cdot \beta^2 \cdot \cos Q}{c \cdot r \cdot (c^2 - \beta^2)}\right)$$

$$\dot{b}_{21} = \mu \cdot \left(\frac{c \cdot \sin Q}{\beta^2 \cdot r} + \frac{Q \cdot \beta^2 \cdot r \cdot \cos Q}{c \cdot (c^2 - \beta^2)}\right)$$

$$b'_{11} = -d\left(\frac{r}{c} - \frac{2 \cdot B_2(\omega_0) \cdot c}{\pi \cdot \beta \cdot r}\right)\sin Q$$

$$\dot{b}'_{12} = \frac{d}{\mu \cdot r}\left(\frac{r}{c} - \frac{2 \cdot B_2(\omega_0) \cdot c}{\pi \cdot \beta \cdot r}\right)\cos Q - \frac{2 \cdot B_2(\omega_0)}{\pi \cdot \beta \cdot r \cdot \rho \cdot \omega}\left(2 - \frac{c^2}{\beta^2 \cdot r^2}\right)\sin Q$$

$$\dot{b}'_{21} = \mu \cdot r \cdot d\left(\frac{r}{c} - \frac{2 \cdot B_2(\omega_0) \cdot c}{\pi \cdot \beta \cdot r}\right)\cos Q + \frac{2 \cdot B_2(\omega_0) \cdot \mu \cdot \beta \cdot r}{\pi \cdot \omega}\left(2 - \frac{c^2}{\beta^2 \cdot r^2}\right)\sin Q$$

Second Case: c < β

We use the here the following notations:

$$r = -i \cdot \left(1 - \left(\frac{c}{\beta}\right)^{1/2}\right)^{1/2} \quad \text{with} \quad r = i \cdot r^*$$

$$Q = \omega \cdot r \cdot \frac{d}{c} = k \cdot r \cdot d \quad \text{with} \quad Q = i \cdot Q^*$$

$$Q^* = \omega \cdot r^* \cdot \frac{d}{c}$$

subsequently:

$$\sin Q = i \cdot \sinh Q^*$$

$$\cos Q = \cosh Q^*.$$

The layer matrix is:

$$b = \begin{bmatrix} \cosh Q^* & \dfrac{\sinh Q^*}{\mu \cdot r^*} \\ -\mu \cdot r^* \sinh Q^* & \cosh Q^* \end{bmatrix}.$$

We have:

$$\dot{b}_{11} = \frac{Q^* \cdot \beta^2 \cdot \sinh Q^*}{c \cdot (c^2 - \beta^2)}$$

$$\dot{b}_{12} = \frac{1}{\mu} \cdot \left(\frac{c \cdot \sinh Q^*}{\beta^2 \cdot r^{*3}} + \frac{Q^* \cdot \beta^2 \cdot \cosh Q^*}{c \cdot r^* \cdot (c^2 - \beta^2)}\right)$$

$$\dot{b}_{21} = \mu \cdot \left(\frac{c \cdot \sinh Q^*}{\beta^2 \cdot r^*} - \frac{Q^* \cdot \beta^2 \cdot r^* \cdot \cosh Q^*}{c \cdot (c^2 - \beta^2)}\right)$$

$$\dot{b}'_{11} = d\left(\frac{r^*}{c} + \frac{2 \cdot B_2(\omega_0) \cdot c}{\pi \cdot \beta \cdot r^*}\right)\sinh Q^*$$

$$\dot{b}'_{12} = \frac{d}{\mu \cdot r^*}\left(\frac{r^*}{c} + \frac{2 \cdot B_2(\omega_0) \cdot c}{\pi \cdot \beta \cdot r^*}\right)\cosh Q^* - \frac{2 \cdot B_2(\omega_0)}{\pi \cdot \beta \cdot r^* \cdot \rho \cdot \omega}\left(2 + \frac{c^2}{\beta^2 \cdot r^{*2}}\right)\sinh Q^*$$

$$\dot{b}'_{21} = -\mu \cdot r^* \cdot d\left(\frac{r^*}{c} + \frac{2 \cdot B_2(\omega_0) \cdot c}{\pi \cdot \beta \cdot r^*}\right)\cosh Q^*$$

$$- \frac{2 \cdot B_2(\omega_0) \cdot \mu \cdot \beta \cdot r^*}{\pi \cdot \omega}\left(2 + \frac{c^2}{\beta^2 \cdot r^{*2}}\right)\sinh Q^*.$$

Third Case: $c = \beta$

Here $r = Q = 0$. Calculations at the limit $c \to \beta$ are required. The layer matrix becomes:

$$B = \begin{bmatrix} 1 & \dfrac{\omega \cdot d}{\mu \cdot c} \\ 0 & 1 \end{bmatrix}.$$

We have:

$$\dot{b}_{11} = -\frac{\omega^2 \cdot d^2}{c^3}$$

$$\dot{b}_{12} = -\frac{\omega \cdot d}{\mu \cdot c^2} \cdot \left(1 + \frac{\omega^2 \cdot d^2}{3 \cdot c^2} \right)$$

$$\dot{b}_{21} = \frac{2 \cdot \mu \cdot \omega \cdot d}{c^2}$$

$$b'_{11} = \frac{2 \cdot d^2 \cdot \omega \cdot B_2(\omega_0)}{\pi \cdot c}$$

$$b'_{12} = \frac{d}{\mu} \cdot \left(\frac{1}{c} - \frac{4 \cdot B_2(\omega_0)}{\pi} \right) + \frac{2 \cdot B_2(\omega_0) \cdot \omega^2 \cdot d^3}{3 \cdot \pi \cdot \mu \cdot c^2}$$

$$b'_{21} = -\frac{4 \cdot \mu \cdot d \cdot B_2(\omega_0)}{\pi}.$$

Appendix B: Computation of the Integral Quantities I_1 and C_2.

The notation *S-L* denotes the transition from the solid to the liquid or rigid halfspace, while the classical solid halfspace will be denoted *S-S*. The dot symbol ($\cdot$) is used here for the derivatives with respect to the time. Let us first define the integral J_m^1, J_m^2 and J_m^3.
With

$$q_m = k \cdot r_{\beta_m} \cdot (z - z_{m-1})$$

we get:

$$J_m^1 = \int_{z_{m-1}}^{z_m} \cos^2 q_m \, dz$$

$$J_m^2 = \int_{z_{m-1}}^{z_m} \sin^2 q_m \, dz$$

$$J_m^3 = \int_{z_{m-1}}^{z_m} \sin q_m \cos q_m \, dz.$$

We have to distinguish three cases: $c > \beta_m$, $c < \beta_m$ and $c = \beta_m$.

For $c > \beta_m$:

$$J_m^1 = \frac{d_m}{2} + \frac{c \cdot \sin(2Q_m)}{4 \cdot \omega \cdot r_{\beta_m}}$$

$$J_m^2 = \frac{d_m}{2} - \frac{c \cdot \sin(2Q_m)}{4 \cdot \omega \cdot r_{\beta_m}}$$

$$J_m^3 = \frac{c \cdot \sin^2 Q_m}{2 \cdot \omega \cdot r_{\beta_m}}$$

$$J_m^4 = \frac{c \cdot \sin^2 Q_m}{2 \cdot \omega \cdot r_{\beta_m}^2} = \frac{J_m^3}{r_{\beta_m}}$$

$$J_m^5 = r_{\beta_m}^2 \cdot J_m^2; \quad J_m^6 = r_{\beta_m}^2 \cdot J_m^4; \quad J_m^7 = \frac{J_m^2}{r_{\beta_m}^2}.$$

For $c < \beta_m$:

$$J_m^1 = \frac{d_m}{2} + \frac{c \cdot \sinh(2Q_m^*)}{4 \cdot \omega \cdot r_{\beta_m}^*}$$

$$J_m^2 = \frac{d_m}{2} - \frac{c \cdot \sinh(2Q_m^*)}{4 \cdot \omega \cdot r_{\beta_m}^*}$$

$$J_m^3 = \frac{i \cdot c \cdot \sinh^2 Q_m^*}{2 \cdot \omega \cdot r_{\beta_m}^*}$$

$$J_m^4 = \frac{c \cdot \sinh^2 Q_m^*}{2 \cdot \omega \cdot r_{\beta_m}^{*2}} = \frac{J_m^3}{r_{\beta_m}}$$

$$J_m^5 = -r_{\beta_m}^{*2} \cdot J_m^2; \quad J_m^6 = -r_{\beta_m}^{*2} \cdot J_m^4; \quad J_m^7 = -\frac{J_m^2}{r_{\beta_m}^{*2}}.$$

For $c = \beta_m$:

$$J_m^1 = d_m; \quad J_m^2 = 0; \quad J_m^3 = 0$$

$$J_m^4 = \frac{\omega \cdot d_m^2}{2c}; \quad J_m^5 = 0; \quad J_m^6 = 0; \quad J_m^7 = \frac{\omega^2 \cdot d_m^3}{3c^2}.$$

B1. Energy Integral

The energy integral is:

$$I_1 = \int_0^\infty \rho \cdot \left(\frac{\dot{v}(z)}{\dot{v}_0}\right)^2 dz.$$

For a layered medium, the energy integral can be written:

$$I_1 = \begin{cases} \left(\dfrac{c}{v_0}\right)^2 \cdot \sum\limits_{m=1}^{n} I_{(m)} & \text{for the } (S-L) \text{ case} \\[2em] \left(\dfrac{c}{v_0}\right)^2 \cdot \left(\left(\sum\limits_{m=1}^{n-1} I_{(m)}\right) + I_{(S-S)}\right) & \text{for the } (S-S) \text{ case} \end{cases} \tag{B1}$$

with:

$$I_{(m)} = \int_{z_{m-1}}^{z_m} \rho_m \cdot \left(\frac{\dot{v}(z)}{c}\right)^2 dz \tag{B2}$$

$$I_{(S-S)} = \int_{z_{n-1}}^{\infty} \rho_n \cdot \left(\frac{\dot{v}(z)}{c}\right)^2 dz. \tag{B3}$$

The integrals can be written:

$$I_{(m)} = -\rho_m \cdot k^2 \cdot \left(v_{m-1}^2 \cdot J_m^1 + \frac{(y_{z_{m-1}})^2 \cdot J_m^7}{k^2 \cdot \mu_m^2} + \frac{2v_{m-1} \cdot y_{z_{m-1}} \cdot J_m^4}{k \cdot \mu_m}\right)$$

$$I_{(S-S)} = \frac{\rho_n \cdot v_{n-1}^2 \cdot k}{2 \cdot r_{\beta_n}^*}$$

where v_m is the displacement and y_{z_m} is the stress at the m-th interface.

B2. Phase Attenuation

The coefficient C_2 is given by:

$$C_2 = \frac{\displaystyle\int_0^{\infty} \mu \cdot B1 \cdot B_2 \cdot \left(\frac{y_z^2}{\mu^2 \cdot k^2} + v^2\right) dz}{c \displaystyle\int_0^{\infty} \mu \cdot v^2 \, dz}.$$

C_2 can also be written:

$$C_2 = \frac{I^1 + I^2}{c \cdot I^3}$$

with:

$$I^1 = -\int_0^{\infty} \mu \cdot B_1 \cdot B_2 \cdot \left(\frac{\dot{v}}{c}\right)^2 dz = k^2 \cdot v_0^2 \cdot \int_0^{\infty} \mu \cdot B_1 \cdot B_2 \cdot \left(\frac{\dot{v}(z)}{v_0}\right)^2 dz$$

$$I^2 = \int_0^{\infty} \mu \cdot B_1 \cdot B_2 \cdot \frac{y_z^2}{\mu^2} \, dz$$

$$I^3 = -\int_0^{\infty} \mu \cdot \left(\frac{\dot{v}}{c}\right)^2 dz = k^2 \cdot v_0^2 \cdot \int_0^{\infty} \mu \cdot \left(\frac{\dot{v}(z)}{v_0}\right)^2 dz.$$

To compute I^1 and I^3 we can use the same scheme as for I_1 in the expression (B1), assuming simple substitution in multiplicative coefficients:

for I^1 we shall use $(\mu_m B_{1_m} B_{2_m} k^2 v_0^2)$ instead of μ_m and
for I^3 we shall use $(\mu_m k^2 v_0^2)$ instead of μ_m.

Let us now consider I^2. For the *S-L* case, we obtain:

$$I^2 = k^2 \cdot \sum_{m=1}^{n} \mu_m \cdot B_{1_m} \cdot B_{2_m} L_m$$

while for the *S-S* case, we have:

$$I^2 = \left(k^2 \cdot \sum_{m=1}^{n-1} \mu_m \cdot B_{1_m} \cdot B_{2_m} L_m \right) - \frac{k \cdot \mu_n \cdot B_{1_n} \cdot B_{2_n} \cdot v_{n-1}^2 \cdot r_{\beta_n}^*}{2}.$$

The quantity L_m is given by:

$$L_m = v_{m-1}^2 \cdot J_m^5 + \frac{y_{z_{m-1}}^2 \cdot J_m^1}{k^2 \cdot \mu_m^2} - \frac{2 \cdot v_{m-1} \cdot y_{z_{m-1}} \cdot J_m^6}{k \cdot \mu_m}.$$

REFERENCES

AKI, K., and RICHARDS, P. G., *Quantitative Seismology* (Freeman and Co., San Francisco, 1980).

BEN-MENAHEM, A. (1961), *Radiation of Seismic Surface Waves from Finite Moving Sources*, Bull. Seismol. Soc. Am. *51*, 401–435.

BEN-MENAHEM, A., and HARKRIDER, D. G. (1964), *Radiation Patterns of Seismic Surface Waves from Buried Dipolar Point Sources in a Flat Stratified Earth*, J. Geophys. Res. *69*, 2605–2620.

CAMPILLO, M., BARD, P.-Y., NICOLLIN, F., and SANCHEZ-SESMA, F. (1988), *The Mexico Earthquake of Septermber 19, 1985—The Incident Wavefield in Mexico City during the Great Michoacan Earthquake and its Interaction with the Deep Basin*, Earthquake Spectra *4* (3), 591–608.

CNEN-ENEL (1977), *Uncorrected Accelerograms. Accelerograms from the Friuli, Italy, Earthquake of May 6, 1976 and Aftershocks: Part 3*, Rome, Italy, November 1977.

DAY, S. M., MCLAUGHLIN, K. L., SKOLLER, B., and STEVENS, J. L. (1989), *Potential Errors in Locked Mode Synthetics for Anelastic Earth Models*, Geophys. Res. Lett. *16*, 203–206.

FÄH, D., SUHADOLC, P., and PANZA, G. F. (1990), *Estimation of Strong Ground Motion in Laterally Heterogeneous Media*, Proc. 9-th European Conf. on Earthquake Eng., Sept. 1990, Moscow, USSR.

FUTTERMAN, W. I. (1962), *Dispersive Body Waves*, J. Geophys. Res. *67*, 5279–5291.

HARKRIDER, D. G. (1970), *Surface Waves in Multilayered Elastic Media. Part II. Higher Mode Spectra and Spectral Ratios from Point Sources in a Plane Layered Earth Model*, Bull. Seismol. Soc. Am. *60*, 1937–1987.

HARVEY, D. J. (1981), *Seismograms Synthesis Using Normal Mode Superposition: The Locked Mode Approximation*, Geophys. J. R. Astr. Soc. *66*, 37–69.

HASKELL, N. A. (1953), *The Dispersion of Surface Waves in Multilayered Media*, Bull. Seismol. Soc. Am. *43*, 17–34.

HEATON, T. H., and HELMBERGER, D. V. (1978), *Predictability of Strong Ground Motion in the Imperial Valley: Modeling the M4.9, November 4, 1976 Brawley Earthquake*, Bull. Seismol. Soc. Am. *68*, 31–48.

KNOPOFF, L. (1964), *A Matrix Method for Elastic Wave Problems*, Bull. Seismol. Soc. Am. *54*, 431–438.

KNOPOFF, L., SCHWAB, F., and KAUSEL, E. (1973), *Interpretation of Lg*, Geophys. J. R. Astr. Soc. *33*, 389–404.

MAO, W. J., and SUHADOLC, P. (1990), *Ray-tracing, Inversion of Travel Times and Waveform Modelling of Strong Motion Data: Application to the Friuli Seismic Area NE Italy*, Geophys. J. Int. Submitted.

MAO, W., SUHADOLC, P., FÄH, D., and PANZA, G. F. (1990), *An Example of Interpretation Strong Motion Data by Source Complexity*, 18th International Conference on Mathematical Geophysics, Jerusalem, Israel (17–22 June 1990).

PANZA, G. F., SCHWAB, F. A., and KNOPOFF, L. (1973), *Multimode Surface Waves for Selected Focal Mechanisms. I. Dip-slip Sources on a Vertical Fault Plane*, Geophys. J. R. Astr. Soc. *34*, 265–278.

PANZA, G. F., and CALCAGNILE, G. (1975), *Lg, Li and Rg from Rayleigh Modes*, Geophys. J. R. Astr. Soc. *40*, 475–487.

PANZA, G. F. (1985), *Synthetic Seismograms: The Rayleigh Waves Modal Summation*, J. Geophys. *58*, 125–145.

PANZA, G. F., and SUHADOLC, P., *Complete strong motion synthetics*. In *Seismic Strong Motion Synthetics* (Bolt, B. A., ed.) (Academic Press, Orlando 1987), Computational Techniques *4*, 153–204.

SCHWAB, F. (1970), *Surface-wave Dispersion Computations: Knopoff's Method*, Bull Seismol. Soc. Am. *60* 1491–1520.

SCHWAB, F., and KNOPOFF, L. (1971), *Surface Waves on Multilayered Anelastic Media*, Bull. Seismol. Soc. Am. *61*, 893–912.

SCHWAB, F., and KNOPOFF, L., *Fast surface wave and free mode computations*, In *Methods in Computational Physics*, vol. 11 (Bolt, B. A., ed.) (New York, Academic Press 1972) pp. 86–180.

SCHWAB, F., and KNOPOFF, L. (1973), *Love Waves and Torsional Free Modes of a Multilayered Anelastic Sphere*, Bull. Seismol. Soc. Am. *63*, 1103–1117.

SCHWAB, F., NAKANISHI, K., CUSCITO, M., PANZA, G. F., LIANG, G., and FREZ, J. (1984), *Surface Wave Computations and the Synthesis of Theoretical Seismograms at High Frequencies*, Bull. Seismol. Soc. Am. *74*, 1555–1578.

SCHWAB, F. (1988), *Mechanism of Anelasticity*, Geophys. J. *95*, 261–284.

SLEJKO, D., and RENNER, G. (1984), In "Finalità ed Esperienze della Rete Sismometrica del Friuli-Venezia Giulia", pp. 75–91. Regione Autonoma Friuli-Venezia Giulia, Trieste.

SUHADOLC, P., CERNOBORI, L., PAZZI, G., and PANZA, G. F., *Synthetic isoseismal: Applications to Italian earthquakes, Seismic Hazard in Mediterranean Regions*, In. J. Bonnin *et al.* (eds.) (Kluwer Academic Press 1988) pp. 205–228.

SWANGER, H. J., and BOORE, D. M. (1978), *Simulation of Strong-Motion Displacements Using Surface-wave Modal Superposition*, Bull. Seismol. Soc. Am. *68*, 907–922.

TAKEUCHI, H., and SAITO, M., *Seismic surface waves*, In *Methods in Computational Physics*, vol. 11, (Bolt, B. A., ed.) (New York, Academic Press 1972) pp. 217–295.

THOMSON, W. T. (1950), *Transmission of Elastic Waves through a Stratified Solid Medium*, J. Appl. Phys. *21*, 89–93.

VACCARI, F., GREGERSEN, S., FURLAN, M., and PANZA, G. F. (1989), *Synthetic Seismograms in Laterally Heterogeneous Anelastic Media by Modal Summation of P-SW-waves*, Geophys. J. Int. *99*, 285–295.

(Received November 20, 1990, revised/accepted May 30, 1991)

PAGEOPH, Vol. 136, No. 4 (1991)

0033–4553/91/040561–16$1.50 + 0.20/0

Body-Wave Dispersion: Measurement and Interpretation

ANTONI M. CORREIG[1,2]

Abstract—Generalizing previous studies on short-period data, it is shown that body-wave dispersion can be measured from broad-band records of earthquakes of moderate magnitude. The method is based on the direct measurement of the arrival time of the frequency components of a seismic wave, and the arrival time is defined by its expectation value. The frequency components of the signal are obtained through a narrow band-pass filtering process. Previous to any interpretation, a correction of the arrival time for instrument response and group delay of the filter is needed. In the first step, body-wave dispersion is related to an absorption band to account for intrinsic attenuation, and thereafter we generalize this interpretation by considering a cascade of filters to account for medium parameters (attenuation and a layered crust) and source parameters (source time function and finiteness of fault). An inversion scheme to obtain the filter parameters can be devised by following, in a formal way, the same procedure as for the case of surface wave dispersion.

Key words: Body waves, attenuation, dispersion.

Introduction

Propagation of seismic waves in an attenuating medium can be modeled as the propagation of seismic waves in an elastic medium convolved with an attenuation operator. The attenuation operator can be phenomenologically described by means of a continuous relaxation model (LIU *et al.*, 1976), defined in terms of short and long relaxation times, a constant Q_m which defines the flat part of the intrinsic attenuation Q-spectrum, and the travel time of the seismic wave. For casuality to be preserved, body waves must be dispersive, attenuation and dispersion being related through the Kramers-Kröning relation.

It has long been accepted that the intrinsic attenuation Q of the medium is a parameter that is practically independent of frequency, and in general so large that body-wave dispersion cannot be measured. However, once the analysis of coda waves for the retrieval of coda Q (AKI and CHOUET, 1975) was widely used, it became apparent that, at high frequencies, coda-Q was strongly frequency dependent, suggesting that intrinsic Q could also be frequency dependent. This result was

[1] Laboratori d'Estudis Geofisics "Eduard Fontsere", IEC, Facultat de Fisica, Marti Franques 1, 08028, Barcelona, Spain.
[2] DGDGP-GEOFISICA, Facultat de Fisica, Marti Franques 1, 08028, Barcelona, Spain.

confirmed by ORCUT (1987), who found a frequency dependent Q for short period shear waves. It has also been found from the analysis of coda waves (see for example CORREIG et al., 1990) that Q may have a very low value locally, of the order of a few tens. Both results (low Q and Q frequency dependence) suggest that body-wave dispersion could be measured. Effectively, body-wave dispersion has been measured by WUENSCHELL (1965) and DOORNBOS (1983) by comparing real data with synthetic seismograms, JACOBSON (1987) through a phase analysis, MCLAUGHLIN and ANDERSON (1987) through an analysis of residual times and HOANG and GRANET (1980) from differential arrival time between P- and S-waves.

CORREIG and MITCHELL (1989) devised a method to measure body-wave dispersion through the measurement of the arrival time of some selected frequency components of the seismogram (and so directly measuring the dispersion) and the method was applied to short period data. The seismograms were band-pass filtered at 1.5, 3, 6, 12 and 24 Hz (the common central frequencies for coda-wave analysis) and the arrival times were picked up "by eye" when they were sharp enough. Once the arrival time of each frequency had been corrected by the group delay of the instrument response and the filter (there was no need for correction for source-time function, because corner frequencies were above 25 Hz), it was clearly seen that the high-frequency components arrived earlier than the low-frequency ones. Finally, the observed P-wave dispersion was interpreted in terms of an absorption band.

It is the aim of this paper to extend the measurement of body-wave dispersion to broad-band data. Firstly, the continuous relaxation model and the transfer function relating the body-wave dispersion to attenuation are introduced, and a definition of arrival time is proposed, independently on the subjectivity of the analyzer (the "by eye" method) and of the noise contents of the filtered signal, so that the same criterion is coherently applied to all readings. Secondly, the filtering process and the group delay of the filter is shown, and the body-wave dispersion is identified with the group delay of a filter. This point of view is generalized in the sense of interpreting the observed group delay as due to a cascade of filters composed by source mechanism and medium parameters. A numerical analysis is performed to find the minimum values of the parameters for which body-wave dispersion can be measured by a broad-band seismometer (with a sampling rate of 8 Hz). Finally, examples of body-wave dispersion measured from records of the NARS array are provided.

SCHWAB (1988) has worked out a scattering mechanism of anelasticity based on the nature of polycrystalline aggregates. From a macroscopic point of view, realistic values of $Q(\omega)$ and $C(\omega)$ should be provided to give accurate estimates of the scattering function. Although a vast set of estimates of the intrinsic Q of the earth exists, practically no data on the dispersive body-wave velocities can be found. This method we present for the measurement of body-wave dispersion may help to fill this gap.

The Measurement of Body-wave Dispersion

CHOY and CORMIER (1986) have shown that the phase shifts due to the propagation and dispersion may be accounted for by the complex travel time $T(\omega)$, the integral of the complex slowness along the ray path:

$$\hat{T}(\omega) = \int \frac{ds}{\hat{c}(\omega)} \tag{1}$$

where $\hat{c}(\omega)$ is the complex, dispersive velocity of a linear viscoelastic solid with no scattering losses and assuming asymptotic ray theory. By imposing causality the following expression (a form of the Kramers–Kröning relation) is found:

$$\operatorname{Re} \hat{T}(\omega) = \operatorname{Re} \hat{T}(\infty) + H[t^*(\omega)]/2 \tag{2}$$

where

$$\operatorname{Re} \hat{T}(\infty) = t = \int \frac{ds}{\operatorname{Re} \hat{c}(\infty)}$$

$$Re \; \hat{T}(\omega) = t(\omega) \tag{3}$$

$$t^*(\omega) = 2 \int \frac{ds}{\operatorname{Im} \hat{c}(\omega)}$$

and H means Hilbert transform. $\operatorname{Re} \hat{T}(\infty)$ represents the fastest arriving frequency component, the infinity frequency, and $\operatorname{Re} \hat{T}(\omega)$ the arrival of frequency component ω.

Equations (2) and (3) suggest that, if some frequency components of the signal are available, the dispersion can be measured as the difference in arrival time of a frequency ω with respect to a reference frequency ω_r:

$$\Delta t = t(\omega) - t(\omega_r). \tag{4}$$

To make the above expression operative, the arrival time must be obtained with very high precision.

For a seismic record in the presence of noise and at large periods, it is difficult to peak up the arrival time with the same criterion at different stations, that is, with different noise contents. Hence, an objective definition of arrival time is needed. The arrival time of a signal (or of a frequency component of a signal) can be defined in the following way (SHIFT, 1949). The expectation value of the arrival time $\langle t \rangle$ of a wave front is given by:

$$\langle t \rangle = \int_{-\infty}^{\infty} \Psi^* t \Psi \, ds \tag{5}$$

where Ψ is the wave function (from a practical point of view, the recorded amplitude) fulfilling the normalization condition

$$\int_{-\infty}^{\infty} \Psi^* \Psi \, dt = 1.$$

(Note that the function $I(t) = |\Psi|^2$ is the probability density of the arrival time of a wave front at a time t.) Similarly, the standard deviation S_t is given by

$$S_t = \int_{-\infty}^{\infty} \Psi^*(t - \langle t \rangle)^2 \Psi \, dt \qquad (6)$$

and the duration D_t of a pulse can be defined as

$$D_t^2 = 2S_t. \qquad (7)$$

Finally, we define the arrival time of the seismic pulse as

$$t_a = \langle t \rangle - D_t / 2. \qquad (8)$$

The above expression will provide us with a way to compute the travel time of the seismic wave and of its frequency components (i.e., the filtered signal). However, as suggested by equation (4), a convenient measure of the dispersion will be the arrival time of the different frequency components of the signal with respect to a frequency taken as a reference. In this paper a frequency of reference of 1 Hz has been considered.

The Continuous Relaxation Model

For a continuous relaxation model (CRM), the internal friction Q^{-1} is expressed as (LIU et al., 1976)

$$Q^{-1}(\omega) = \frac{2}{\pi} Q_m^{-1} \arctan \left[\frac{\omega(\tau_1 - \tau_2)}{1 + \omega^2 \tau_1 \tau_2} \right] \qquad (9)$$

where ω is the angular frequency and τ_1 and τ_2 are the long and short relaxation times respectively, $\tau_1 = s_1^{-1}$ and $\tau_2 = s_2^{-1}$, s_1 and s_2 are the low and high cutoff frequencies. Q_m^{-1} is a constant that defines the flat part of the $Q^{-1}(\omega)$ spectrum, bounded by $s_1 < \omega < s_2$. This model is independent of the physical mechanism that produces the loss of energy in the medium.

In an attenuative medium, if causality has to be preserved, body waves must be dispersive. Attenuation and dispersion are related through Kramers–Kröning equations. By modeling the attenuative medium as a linear viscoelastic solid, the phase velocity is expressed as (BEN-MENAHEM and SINGH, 1981)

$$C(\omega) = C_\infty \left(1 + \frac{1}{\pi} Q_m^{-1} \left[\ln \frac{\tau_1}{\tau_2} + \frac{1}{2} \ln \left(\frac{1 + \omega^2 \tau_2^2}{1 + \omega^2 \tau_1^2} \right) \right] \right)^{-1} \qquad (10)$$

where C_∞ and $C(\omega)$ are the phase velocities at infinite frequency and at frequency ω, respectively. The other parameters are as in equation (9).

If attenuation is large enough, Δt can be observed and related to model parameters through equations (2) and (4). It can be shown (CORREIG, 1991) that the transfer function that relates dispersion to attenuation can be written as

$$\Delta t = \frac{t}{2\pi} Q_m^{-1} \ln\left[\frac{(1+\omega^2\tau_2^2)(1+\omega_r^2\tau_1^2)}{(1+\omega^2\tau_1^2)(1+\omega_r^2\tau_2^2)}\right] \tag{11}$$

where t is the travel time.

Figure 1 shows phase velocity $C(\omega)/C(\infty)$ (obtained from equation (10)) and group velocity (obtained through numerical derivative of phase velocity) for two models, one defined by $Q_m = 200$, $\tau_1 = 300$ s and $\tau_2 = 0.001$ s, called MINSTER (MINSTER, 1978), and the second defined by $Q_m = 5$, $\tau_1 = 43$ and $\tau_2 = 0.37$, called CERDANYA (CORREIG and MITCHELL, 1989). Whereas body-wave dispersion could hardly be measured for the former model, it was obtained without any difficulty for the latter.

Data Analysis

The seismic phase to be analyzed (for example a P- PKP-, S-wave pulse, or any other of interest) is band-pass filtered using a 6-pole Butterworth filter centered at the frequencies of interest and with a bandwidth that is 2/3 of the central frequency. Ideally we would like to measure the arrival time of a monochromatic wave or at

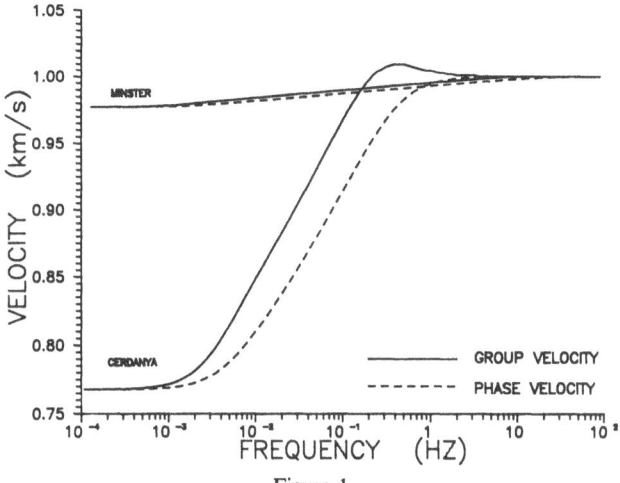

Figure 1

Phase and group velocity dispersion curves for two attenuating media, CERDANYA, with a very low Q_m and close relaxation times (i.e., with a very large attenuation), and MINSTER with a moderate value of Q_m and very separated relaxation times. For more explanation see the text.

least a very narrow band of frequencies. However, as we take narrower filters, numerical instabilities appear and it is harder to control the group delay of the filter. A bandwidth of 2/3 of the central frequency appears to satisfy our requirements.

As is well known, the filtering process implies a group delay of the signal, so that before interpreting the differences in arrival time in terms of dispersion the readings must be corrected for this group delay. The group delay $\tau(\omega)$ of a filter is defined as (SEIDL and STAMMLER, 1984)

$$\tau(\omega) = -d\phi(\omega)/d\omega, \tag{12}$$

where $\phi(\omega)$ is the phase of the filter. The mean group delay of the filter is defined by:

$$\tau_m = \frac{\displaystyle\int_0^\infty \tau(\omega)|H(i\omega)|^2 \, d\omega}{\displaystyle\int_0^\infty |H(i\omega)|^2 \, d\omega} \tag{13}$$

where $H(i\omega)$ is the transfer function of the filter.

Figure 2 shows the group delay of a NARS station (DOST, 1987) as computed from the poles and zeros of the transfer function (such as appear in the header of the CD ROM with digital data as distributed by NEIC and ORFEUS). The unfiltered seismogram will be delayed by the mean group delay of the instrument, 0.22 s, whereas the filtered signal will be affected by (approximately) the instantaneous group delay of the instrument corresponding to the central frequency of the

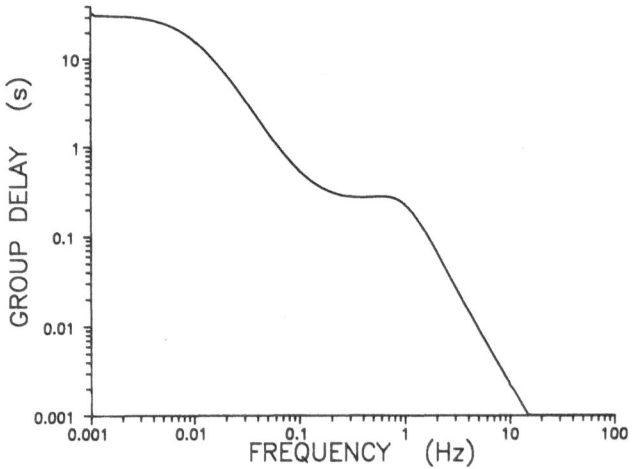

Figure 2
Instrumental group delay of a NARS station.

band-pass filter. The instantaneous group delay of the instrument is 0.207, 0.319 and 0.508 s for corresponding periods of 1, 5 and 10 s.

Figure 3 shows the amplitude, phase response and group delay of the 6-pole Butterworth filter centered at periods of 1, 2, 5, and 10 s. The corresponding group delay of the filter (taken as the mean group delay of each filter, quite close to the instantaneous of the central frequency) is 0.136, 0.672 and 1.343 s for the corresponding periods of 1, 5 and 10 s.

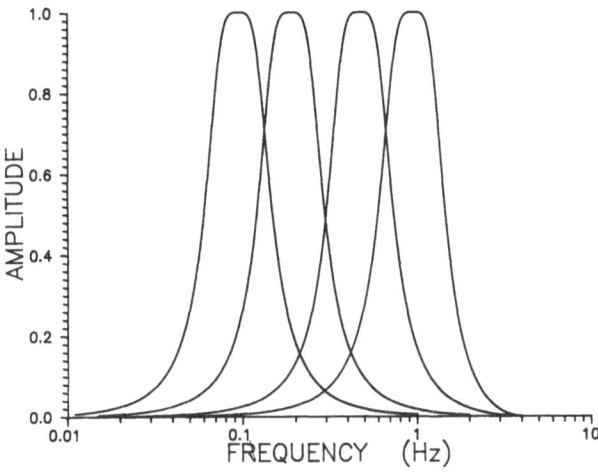

Figure 3a.

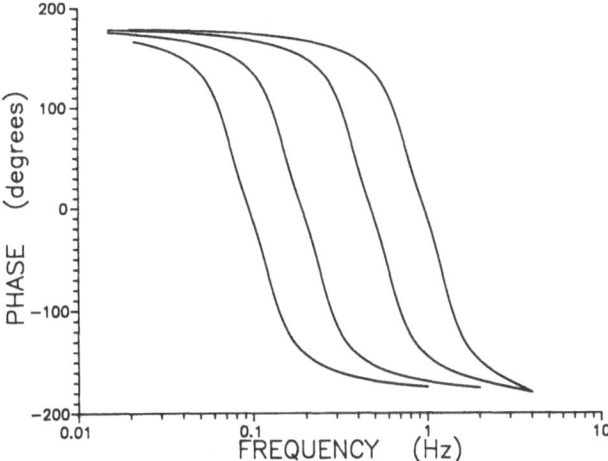

Figure 3b

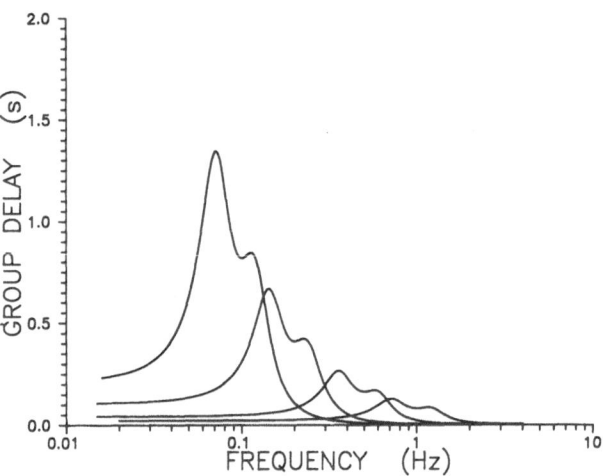

Figure 3c.

Figure 3

Amplitude, phase and group delay of 6-pole Butterworth band-pass filters centered at frequencies of 0.1, 0.2, 0.5 and 1 Hz with a band with 2/3 of the central frequency.

The Medium and the Source as a Cascade of Filters

The measured dispersion can be interpreted as being due to the group delay of a cascade of filters arising from the source and the medium. From the point of view of the medium the main contributions will be due to attenuation and crustal transfer function, and for the source to time function and finiteness. Because of linearity, the resulting group delay will be the sum of the group delay of each filter. In the next section we will analyze the various contributions.

Attenuation Operator

To interpret the dissipation of the medium as a filter we can define an attenuation operator $D(\omega)$. Attenuation operators (MÜLLER, 1983) are wave profiles $u(x, t)$ of a plane wave propagating in x-direction, where the input at $x = 0$ corresponds to the delta function:

$$u(x, t) = \frac{1}{2\pi} \int_{-\infty}^{\infty} \exp\left[i\omega\left(t - \frac{x}{\hat{c}(\omega)} \right)\right] d\omega \tag{14}$$

where $\hat{c}(\omega)$ is the complex wave velocity. Equation (14) can be rewritten as

$$u(x, t) = \frac{1}{2\pi} \int_{-\infty}^{\infty} D(\omega)\exp\left[i\omega\left(t - \frac{x}{C(\omega)} \right)\right] d\omega, \tag{15}$$

where $C(\omega)$ is the phase velocity. The attenuation operator $D(\omega)$ is given by (MINSTER, 1978)

$$D(\omega) = \exp\left(i\omega t_u \left\{\frac{\tau}{t_u} + 1 - \left[1 + \frac{2}{\pi Q_m}\ln\frac{i\omega + \tau_1^{-1}}{i\omega + \tau_2^{-1}}\right]^{-1/2}\right\}\right) \qquad (16)$$

where t_u is the travel time and $\tau = t - t_u$. The corresponding group delay will be obtained by taking the derivative of its phase with respect to ω. (Of course the group delay computed through (12) coincides with the dispersion computed from (11). The operator defined by (16) is very useful because it can be convolved with the elastic response of the medium.)

Figure 4 shows the variation of the group delay (GD) of the attenuation operator as a function of the variation of its parameters. As a reference, the model

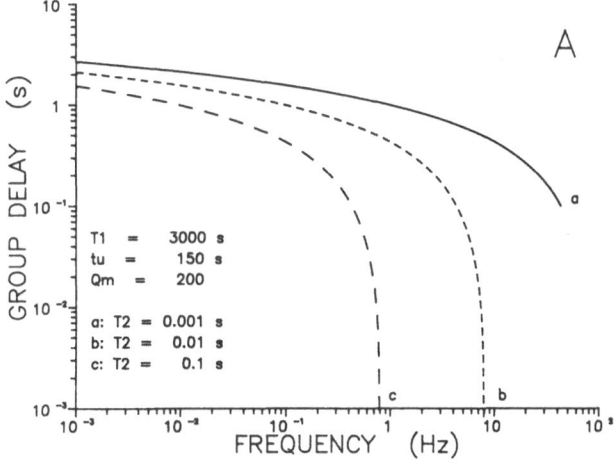

Figure 4a

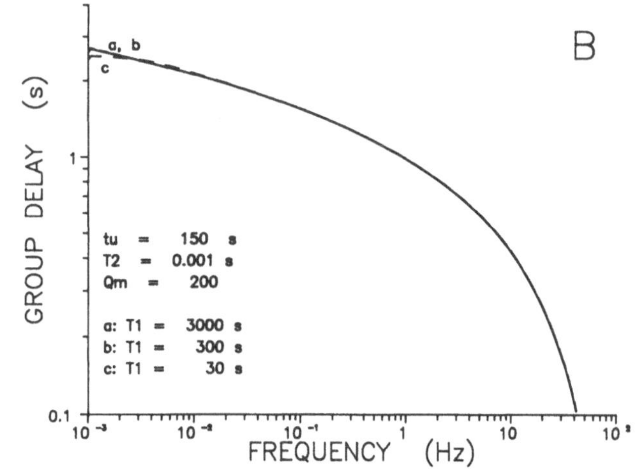

Figure 4b

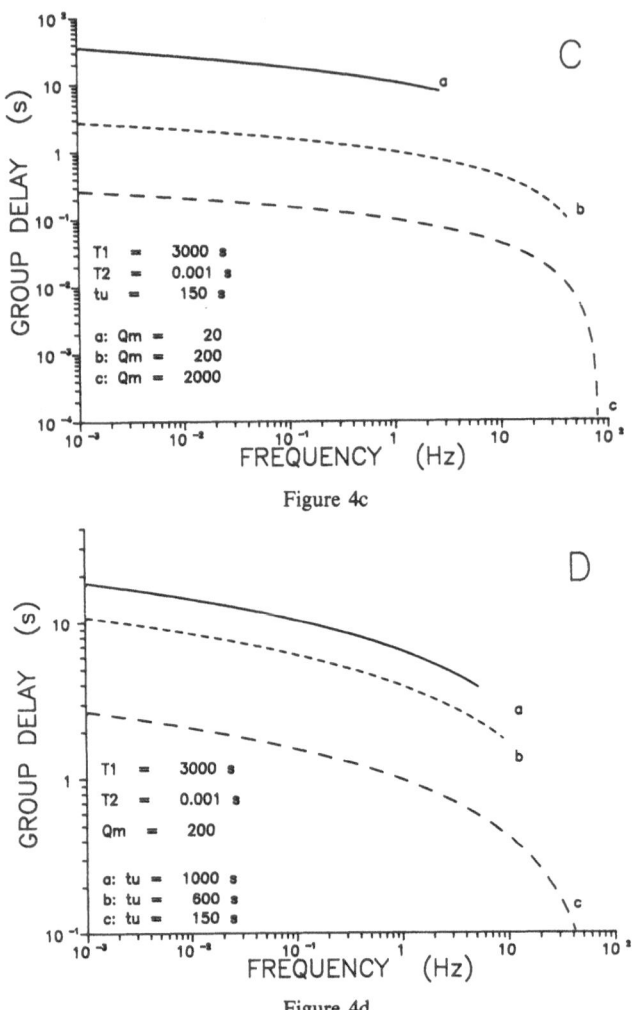

Figure 4c

Figure 4d

Figure 4

Variation of the group delay of the attenuation operator filter as a function of (A) the short-term relaxation time, (B) the long-term relaxation time, (C), the Q_m parameter and (D) the travel time of the wave. The group delay is sensitive to all parameters excepting the large term relaxation time.

parameters analyzed by MINSTER (1978) has been used. In **A** the variation of the *GD* with respect to the short relaxation time τ_1 is shown; at high frequencies the model is quite sensitive to its variation. In **B** the variation of *GD* with respect to the long relaxation time is represented; as can be seen, the model is insensitive to this parameter. In **C** the variation of the *GD* as a function of Q_m is shown to be very large, as in **D** where the variation of *GD* with the travel time is represented. In all cases the *GD* is a decreasing function of frequency, and can be discriminated by a broad-band seismometer. A more rigorous analysis, if needed, can be performed through the partial derivatives of Q with respect to the model parameter (that can be obtained analytically, for example from equation (9)).

Source Time Function

Consider, for simplicity, the exponential decay function

$$s(t) = H(t)e^{-t/\tau}, \ \tau > 0 \tag{17}$$

where τ is the rise time. Its Fourier transform is

$$S(\omega) = \frac{\tau}{1 + i\omega\tau} \tag{18}$$

and its corresponding group delay

$$r(\omega) = \frac{\tau}{1 + (\omega\tau)^2}. \tag{19}$$

Figure 5 shows the group delay of the exponential decay function for rise times of 0.10 and 0.25 s. In both cases the group delay is significant and can be measured without difficulty if observations are made at periods that cover the inflection of the flat part to the decaying part of the spectrum. A rise time of 0.25 s was found by CORREIG and MITCHELL (1980) for earthquakes of magnitude $M_s = 5$ occurred at the east Pacific rise.

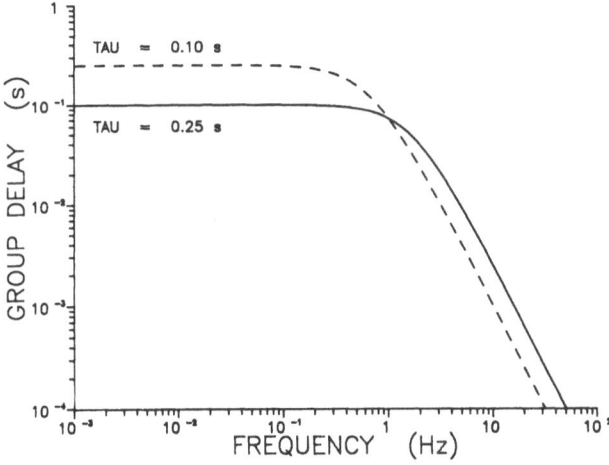

Figure 5

Variation of the group delay of a decaying exponential function (representing a source time function) as a function of the rise time.

Finiteness of the Source

The finiteness of the source can be modeled as (BEN-MENAHEM and SINGH, 1981)

$$f(\omega) = \frac{\sin X}{X} e^{-iX} \tag{20}$$

$$X(\omega) = \omega \frac{L}{2v} \left(1 - \frac{v}{c} \cos \Theta \right)$$

where L is the fault length, v the rupture velocity, c the wave velocity and Θ the angle between the direction of rupture and the epicenter-receiver direction. Because the phase of this filter is linear with ω, the corresponding group delay will be constant, for a given azimuth, for the whole spectrum, and will take different values for different azimuths. This group delay can thus be obtained from observations at distinct azimuths.

Spectral Response of a Multilayered Crust

The influence of a layered crust on the amplitude and phase of body waves has long been recognized (PHINNEY, 1964). By interpreting the crustal transfer function as a filter, its corresponding group delay can be computed. The amplitude and phase of the crustal transfer function have been computed through the Haskell matrix formalism (HASKELL, 1953; BEN-MENAHEM and SINGH, 1981). Figure 6 shows the group delay to two models, an average continental crust 37 km thick and

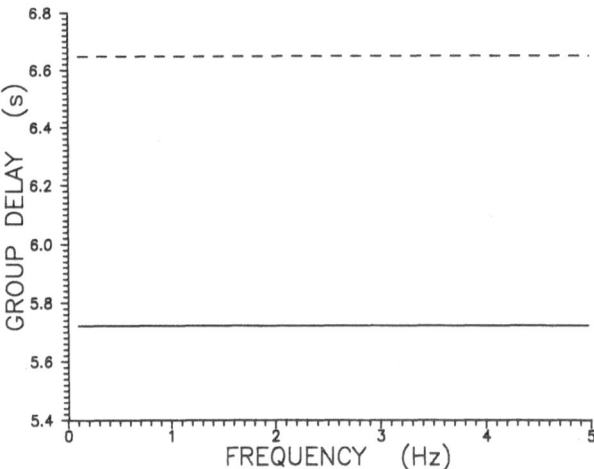

Figure 6
Group delay of the crustal transfer function for a crust 37 km thick (solid line) and 49 km thick (dashed line). This filter presents no dispersion.

an extreme continental crust 49 km thick. In no case can the dispersion be appreciated, so there is no need for crustal transfer function correction. Clearly, the (constant) group delay measured corresponds to the travel time of the wave across the crust. Note that this group delay has been computed as an effect of the layering only, and so an elastic stack of layers has been considered. The intrinsic attenuation of a real crust will be included in the global measurement of Q through the path source-receiver.

Application and Discussion

Body-wave dispersion has been computed for the events recorded by the broad-band NARS array previously studied by DOST (1987). The reason for the selection of these events is that they are well aligned with the NARS stations, so that there is no need to correct by azimuthal variations of radiation pattern. Characteristics of the array and events are given by DOST (1987). Figure 7 shows the location of the events and the array.

Figure 8 shows the observed body-wave dispersion for events 24/03/84, located at Kurile Islands and hereafter named event N, and 22/12/83, located at NW Africa and hereafter named event S. Records were band-pass filtered by using 6-pole Butterworth filters centered at 0.5, 1, 2, 3, 4, 5, 6, 8 and 10 seconds and with a half-band of $1/3T$ s. Data were corrected by instrumental response and filter group delay. As a reference, the arrival time of the signal filtered at 1 s was used, so that the dispersion is presented as the arrival time of the signal filtered at a period T minus the arrival time of the signal filtered at 1 s. Although there is some scatter, for event N the dispersion presents a peak centered at a period of 5–6 s that ranges from 0.3 to 0.6 s (as a comparison, in a study on the earth's absorption band Doornbos found a dispersion ranging from 0.4 to 1.7 s); at longer periods the dispersion tends to decrease.

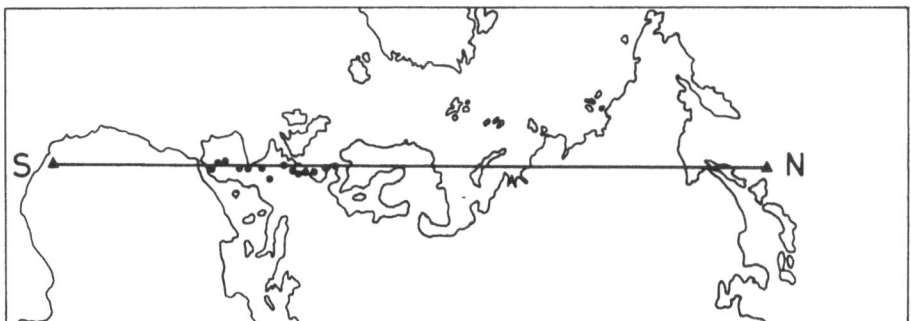

Figure 7

Location of the events and seismic stations (NARS array) used in this study. Redrawn from DOST (1987).

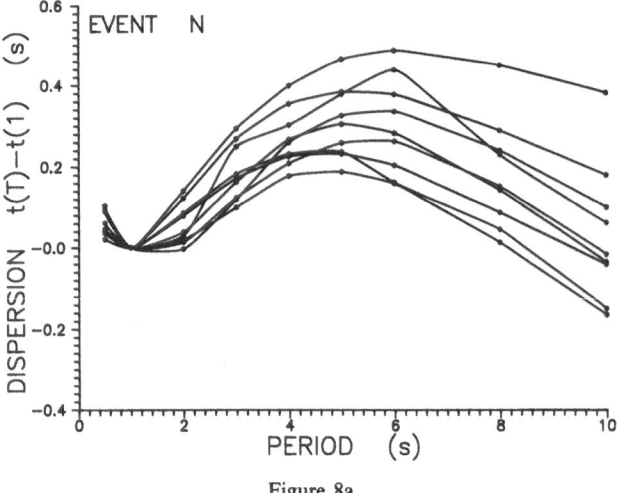

Figure 8a

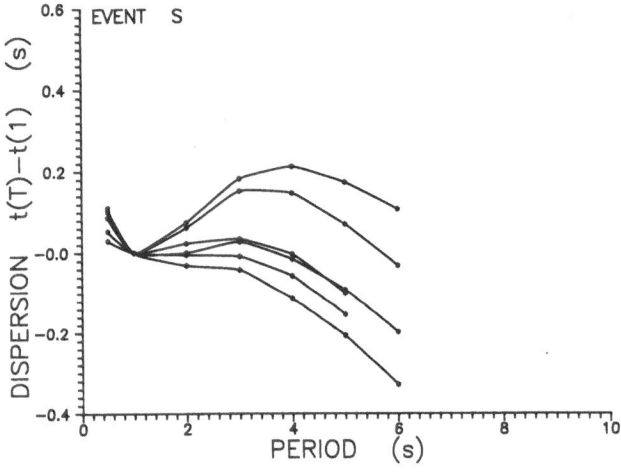

Figure 8
Observed body-wave dispersion for (a) event 24/03/84 located at Kurile Islands and (b) event 22/12/83,
located at NW Africa. For more explanation see the text.

For event *S*, the peak is centered at 3–4 s and ranges from 0.05 to 0.2 s,
although for two stations there is no peak and the dispersion is monotonically
decreasing; at long periods (the longer period that it has been possible to measure
is 6 s for this event) the dispersion tends to be inverse for all events, telling us that
long periods arrive later than the reference period of 1 s, i.e., inverse dispersion. The
differences in the shape of the dispersion curves of the two events can be interpreted
in terms of the different medium characteristics crossed by the waves as well as
distinct source time functions, which will present a large complexity than that
presented by the exponential decay. The presence of the peak is harder to explain

unless we take into account, i) the superposition of several absorption bands with different values of the cutoff frequencies and alternate values of Q_m, a mixture of high and low values, and ii) the scattering of the body-waves, not contemplated in the present study and that can be prominent at short periods. The study of both effects will be the subject of a later paper.

In conclusion, body-wave dispersion can be measured and interpreted in terms of a cascade of filters in which the body-wave dispersion corresponds to the group delay of the filters.

Acknowledgements

This research was supported in part by the DGICYT under Grant PB86–0431–C05–03. The author would like to acknowledge the suggestions from an anonymous referee, which improved the original manuscript.

REFERENCES

AKI, K., and CHOUET, B. (1975), *Origin of Coda Waves: Source, Attenuation and Scattering Effects*, J. Geophys, Res. *80*, 3322–3342.

BEN-MENAHEM, A., and SINGH, J. S., *Seismic Waves and Sources* (Springer-Verlag, New York 1981).

CHOY, G. L., and CORMIER, V. F. (1986), *Direct Measurements of the Mantle Attenuation Operator from Broadband P and S Waveforms*, J. Geophys. Res. *91*, 7326–7342.

CORREIG, A. M. (1991), *On the Measurement of Body Wave Dispersion*, J. Geophys. Res. (In press).

CORREIG, A. M., and MITCHELL, B. J. (1980), *Regional Varitions of Rayleigh Wave Attenuation Coefficients in the Eastern Pacific*, Pure Appl. Geophys. *118*, 831–845.

CORREIG, A. M., and MITCHELL, B. J. (1989), *Attenuative Body-wave Dispersion at La Cerdanya, Eastern Pyrenees*, Phys. Earth Planet. Int. *57*, 304–310.

CORREIG, A. M., MITCHELL, B. J., and ORTIZ, R. (1990), *Seismicity and Coda Q Values in the Eastern Pyrenees: Results from the La Cerdanya Seismic Newtwork*, Pure Appl. Geophys. *132*, 312–329.

DOORNBOS, D. J. (1983), *Observable Effects of the Seismic Absorption Band in the Earth*, Geophys. J. R. Astr. Soc. *75*, 693–711.

DOST, B. (1987), *The NARS Array, a Seismic Experiment in Western Europe*, Ph.D. Thesis, Utrecht University, the Netherlands.

HASKELL, N. A. (1953), *The Dispersion of Surface Waves on a Multilayered Media*, Bull. Seismol. Soc. Am. *43*, 17–34.

HOANG TRONG, P., and GRANET, M. (1980), *Body-wave Dispersion in the Friuli Focal Area*, Phys. Earth Planet. Int. *21*, 31–37.

JACOBSON, R. S. (1987), *An Investigation into the Fundamental Relationship between Attenuation, Phase Dispersion and Frequency, Using Seismic Refraction Profiles over Sedimentary Structures*, Geophys. *52*, 72–87.

LIU, H. P., ANDERSON, D. L., and KANAMORI, H. (1976), *Velocity Dispersion due to Anelasticity: Implications for Seismology and Mantle Composition*, Geophys. J. R. Astr. Soc. *89*, 933–964.

MCLAUGHLIN, K. L., and ANDERSON, L. M. (1987), *Stochastic Dispersion of Short-period P-waves due to Scattering and Multipathing*, Geophys. J. R. Astr. Soc. *89*, 933–964.

MINSTER, J. B. (1978), *Transient and Impulse Responses of a One-dimensional Linearly Attenuating Medium — II. A Parametric Study*, Geophys. J. R. Astr. Soc. *52*, 503–524.

MÜLLER, G. (1983), *Rheological Properties and Velocity Dispersion of a Medium with Power-law Dependence of Q on Frequency*, J. Geophys. *54*, 20–29.

ORCUTT, J. A. (1987), *Structure of the Earth's Oceanic Crust and Uppermost Mantle*, Rev. Geophys. *25*, 1177–1196.

PHINNEY, R. A. (1964), *Structure of the Earth's Crust from Spectral Behavior of Long-period Body-waves*, J. Geophys. Res. *69*, 2997–3017.

SCHWAB, F. (1988), *Mechanism of Anelasticity*, Geophys. J. *95*, 261–284.

SEIDL, D., and STAMMLER, W. (1984), *Restoration of Broad-band Seismograms (Part I)*, J. Geophys. *54*, 114–122.

SHIFT, L. I., *Quantum Mechanics* (McGraw Hill 1949) 404 pp.

WUENSCHEL, P. C. (1965), *Dispersive Body Waves. An Experimental Study*, Geophys. *30*, 539–551.

(Received January 18, 1991, revised/accepted May 21, 1991)

Reprints from PAGEOPH

Aspects of Pacific Seismicity

Edited by
E.A. Okal

1991. 200 pages. Softcover
sFr. 68.– / DM 78.–
ISBN 3-7643-2589-5

Deep Earth
Electrical Conductivity

Edited by
W.H. Campbell

1990. 96 pages. Softcover
sFr. 49.– / DM 58.–
ISBN 3-7643-2564-X

Earthquake Hydrology and
Chemistry

Edited by
C.-Y. King

1985. 478 pages. Softcover
sFr. 84.– / DM 101.–
ISBN 3-7643-1743-4

Earthquake Prediction

Edited by
K. Shimazaki and W.D. Stuart

1985. 238 pages. Softcover
sFr. 70.– / DM 84.–
ISBN 3-7643-1742-6

Electrical Properties
of the Earth's Mantle

Edited by
W.H. Campbell

1987. 498 pages. Softcover
sFr. 52.– / DM 62.–
ISBN 3-7643-1901-1

Fractals in Geophysics

Edited by
C.H. Scholz and B.B. Mandelbrot

1989. 313 pages. Softcover
sFr. 78.– / DM 94.–
ISBN 3-7643-2206-3

Instabilities in Continuous Media

Edited by
*L. von Knopoff, V.I. Keilis-Borok
and G. Puppi*

1985. 216 pages. Softcover
sFr. 82.– / DM 98.–
ISBN 3-7643-1704-3

Intermediate-Term
Earthquake Prediction

Edited by
W.D. Stuart and K. Aki

1988. 718 pages. Softcover
sFr. 86.– / DM 98.–
ISBN 3-7643-1978-X

Ionospheric Modeling

Edited by
J.N. Korenkov

1988. 376 pages. Softcover
sFr. 62.– / DM 74.–
ISBN 3-7643-1926-7

Middle Atmosphere

Edited by
A.R. Plumb and R.A. Vincent

1989. 472 pages. Softcover
sFr. 78.– / DM 88.–
ISBN 3-7643-2290-X

Oceanic Hydraulics

Edited by
R.L. Hughes

1990. 184 pages. Softcover
sFr. 48.– / DM 56.–
ISBN 3-7643-2498-8

Physics of Fracturing and Seismic Energy Release

Edited by
J. Kozak and L. Waniek

1987. 974 pages. Softcover
sFr. 58.– / DM 68.–
ISBN 3-7643-1863-5

Quiet Daily Geomagnetic Fields

Edited by
W. H. Campbell

1989. 244 pages. Softcover
sFr. 64.– / DM 74.–
ISBN 3-7643-2338-8

The Radiosonde Intercomparison SONDEX

Edited by
H. von Richner and P.D. Philips

1984. 352 pages. Softcover
sFr. 108.– / DM 130.–
ISBN 3-7643-1614-4

Scattering and Attenuation of Seismic Waves

Edited by
R.-S. Wu and K. Aki

Part I:
1988. 448 pages. Softcover
sFr. 78.– / DM 94.–
ISBN 3-7643-2254-3

Part II:
1989. 198 pages. Softcover
sFr. 49.80 / DM 59.80
ISBN 3-7643-2341-8

Part III:
1990. 438 pages. Softcover
sFr. 102.– / DM 122.–
ISBN 3-7643-2342-6

Seismicity in Mines

Edited by
S.J. Gibowicz

1989. 680 pages. Softcover
sFr. 66.– / DM 79.80
ISBN 3-7643-2273-X

Subduction Zones

Edited by
L.J. Ruff and H. Kanamori

Part I:
1988. 352 pages. Softcover
sFr. 62.– / DM 74.–
ISBN 3-7643-1928-3

Part II:
1989. 376 pages. Softcover
sFr. 52.– / DM 62.–
ISBN 3-7643-2272-1

Weather and Weather Maps
A Volume Dedicated to the Memory of Tor Bergeron

Edited by
G.H. Liljequist

1981. 292 pages. Softcover
sFr. 97.– / DM 116.–
ISBN 3-7643-1192-4

All volumes available. Please order through your bookseller or directly from the publisher.